# Undergraduate Topics in Computer Science

'Undergraduate Topics in Computer Science' (UTiCS) delivers high-quality instructional content for undergraduates studying in all areas of computing and information science. From core foundational and theoretical material to final-year topics and applications, UTiCS books take a fresh, concise, and modern approach and are ideal for self-study or for a one- or two-semester course. The texts are all authored by established experts in their fields, reviewed by an international advisory board, and contain numerous examples and problems, many of which include fully worked solutions.

More information about this series at http://www.springer.com/series/7592

Gerard O'Regan

# Concise Guide to Software Testing

Gerard O'Regan
SQC Consulting
Mallow, Cork, Ireland

ISSN 1863-7310 ISSN 2197-1781 (electronic)
Undergraduate Topics in Computer Science
ISBN 978-3-030-28493-0 ISBN 978-3-030-28494-7 (eBook)
https://doi.org/10.1007/978-3-030-28494-7



This Springer imprint is published by the registered company Springer Nature Switzerland AG
The registered company address is: Gewerbestrasse 11, 6330 Cham, Switzerland

*To*

*Noel, Maura and Ger Forde*
*(For friendship)*

# Preface

## Overview

The objective of this book is to provide a concise introduction to the software testing field to students and practitioners. The principles of software testing are discussed, and the goal is to give the reader a grasp of its fundamentals, as well as guidance on applying the theory in an industrial environment.

## Organization and Features

Chapter 1 discusses fundamentals of the software quality field, including a history of quality and the pioneers who have influenced the field. We discuss the contributions of Deming, Juran, and Crosby, as well as the work of Watts Humphrey, who is considered the father of software quality.

Chapter 2 presents the fundamentals of software engineering, and we discuss various software lifecycles used in software development. We discuss the various activities in the waterfall model, including requirements gathering and specification, software design, implementation, testing, and maintenance. The lightweight Agile methodology has become very popular in industry replacing the traditional waterfall lifecycle model.

Chapter 3 discusses the fundamentals of testing in traditional software engineering, and we discuss the various types of testing that may be carried out during the project. We discuss test planning, test case definition, test environment set-up, test execution, test tracking, test metrics, test reporting, and testing in an e-commerce environment.

Chapter 4 discusses static testing, which plays an important role in building quality into a product. We discuss the well-known Fagan inspection process developed at IBM in the 1970s, as well as lighter review and walkthrough methodologies. We discuss the static code analysis of software code without executing the code, which is performed with automated tools.

Chapter 5 discusses software test planning, which involves defining the scope of the testing to be performed; defining the test environment; estimating the effort required to define the test cases and to perform the testing; identifying the resources needed (including people, hardware, software, and tools); assigning the resources to the tasks; defining the schedule; and identifying any risks to the testing and managing them.

Chapter 6 discusses test analysis and design, which is concerned with determining the test conditions, and designing the test cases (using various techniques) for the testing. The requirements and test conditions are used to specify the test cases, where each test case includes test input, the procedure for carrying out the test, and the expected results.

Chapter 7 discusses test management, which is concerned with the activities involved in managing the software testing process. A well-planned and managed testing process enables teams to deliver high-quality products on time and on budget. A good test process is repeatable and predictable.

Chapter 8 is concerned with test outsourcing, which involves the selection and management of a testing supplier. It discusses how candidate suppliers are identified, formally evaluated against defined selection criteria, and how the appropriate supplier is selected. The selected supplier is then managed during the testing.

Chapter 9 is concerned with test metrics and problem-solving, and we discuss the well-known goal, question, metrics (GQM) approach. This enables metrics related to the organization goals to be defined. A selection of metrics is presented, and several problem-solving tools such as fishbone diagrams, Pareto charts, trend charts are discussed.

Chapter 10 discusses tools to support various software testing activities. The focus is first to define the process and then to find tools to support the process. Tools to support test management are discussed, as well as tools to support test design and execution. Finally, we discuss tools for static testing, performance and monitoring tools, and defect tracking tools.

Chapter 11 discusses test process improvement and begins with a discussion of a software process and the benefits that may be gained from a software process improvement initiative. We discuss the Capability Maturity Model Integration (CMMI) and dedicated test process improvement models such as the TMM, TPI, TMap, STEP, and CTP.

Chapter 12 discusses testing in the Agile world. Agile is a popular lightweight approach to software development, and it provides opportunities to assess the direction of a project regularly throughout the development lifecycle. Ongoing changes to requirements are considered normal in the Agile world, and Agile has a strong collaborative style of working. It advocates adaptive planning and evolutionary development.

Chapter 13 discusses the verification of safety-critical systems, where such systems often need an extra level of assurance in their correctness. Formal methods support the development and verification of safety-critical systems, and they consist of a set of practical mathematical techniques. They may be employed to rigorously state the requirements of the proposed system and to derive a program from its mathematical specification. Further, they may be employed to provide a rigorous proof that the implemented program satisfies its specification.

Chapter 14 discusses legal and ethical aspects of software testing, and we discuss professional responsibilities of the professional tester. We discuss legal aspects of failure including lawsuits and the law of tort.

Chapter 15 discusses software configuration management and discusses the fundamental concept of a baseline. Configuration management is concerned with identifying those deliverables that must be subject to change control and controlling changes to them. We discuss the importance of configuration management in the testing field.

Chapter 16 is the concluding chapter in which we summarize the journey that we have travelled in this book.

## Audience

The main audience of this book are computer science students who are interested in learning about software testing and in learning on how to build high-quality and reliable software on time and on budget. It will also be of interest to industrialists including software engineers, software testers, quality professionals and software managers, as well as the motivated general reader.

## Acknowledgements

I am deeply indebted to family and friends who supported my efforts in this endeavour and my thanks, as always, to the team at Springer. A special thanks to the Fordes for their friendship and camaraderie over many years (Noel for leaving me in a cloud of dust during the An Oige cycles; Ger for fun and entertainment; and Maura for being a voice of reason and keeping everything together).

Cork, Ireland Gerard O'Regan

# Contents


**1 Fundamentals of Software Quality** . . . 1
1.1 Introduction . . . 1
1.2 History of Software Failures . . . 3
1.3 Background to Software Quality . . . 4
1.3.1 What Is Software Quality? . . . 5
1.3.2 Early Quality Management . . . 5
1.3.3 Total Quality Management . . . 6
1.3.4 Software Quality Control . . . 6
1.4 History of Quality . . . 7
1.4.1 Shewhart . . . 7
1.4.2 Deming . . . 8
1.4.3 Juran . . . 10
1.4.4 Crosby . . . 12
1.4.5 Watts Humphrey . . . 16
1.4.6 Miscellaneous Quality Gurus . . . 18
1.5 Modern Software Quality Management . . . 19
1.5.1 Software Inspections . . . 19
1.5.2 Software Testing . . . 20
1.5.3 Software Quality Assurance . . . 20
1.5.4 Problem-Solving Techniques . . . 21
1.5.5 Cost of Quality . . . 22
1.5.6 Software Process Improvement . . . 23
1.5.7 Software Metrics . . . 24
1.5.8 Customer Satisfaction . . . 25
1.5.9 Assessments (Appraisals) . . . 26
1.5.10 Total Quality Management . . . 27
1.6 Miscellaneous . . . 28
1.6.1 Organization Culture and Change . . . 28
1.6.2 Law of Negligence . . . 29
1.6.3 Quality and the Web . . . 30
1.7 Review Questions . . . 30
1.8 Summary . . . 30
References . . . 31

2 **Fundamentals of Software Engineering** 33
2.1 Introduction 33
2.2 What Is Software Engineering? 36
2.3 Challenges in Software Engineering 39
2.4 Software Processes and Lifecycles 40
2.4.1 Waterfall Lifecycle 41
2.4.2 Spiral Lifecycles 41
2.4.3 Rational Unified Process 43
2.4.4 Agile Development 44
2.5 Activities in Waterfall Lifecycle 46
2.5.1 User Requirements Definition 46
2.5.2 Specification of System Requirements 47
2.5.3 Design 48
2.5.4 Implementation 49
2.5.5 Software Testing 49
2.5.6 Support and Maintenance 51
2.6 Software Inspections 52
2.7 Software Project Management 52
2.8 CMMI Maturity Model 53
2.9 Formal Methods 54
2.10 Review Questions 55
2.11 Summary 55
References 56

3 **Fundamentals of Software Testing** 59
3.1 Introduction 59
3.2 Software Test Process 61
3.3 Software Test Planning and Scheduling 65
3.4 Test Case Design and Definition 66
3.5 Test Execution 67
3.6 Test Reporting and Project Sign-off 67
3.7 Testing and Quality 68
3.7.1 What Is a Software Defect? 68
3.7.2 Is Exhaustive Testing Possible? 68
3.7.3 How Much Testing Should Be Done? 69
3.7.4 Testing and Quality Improvement 69
3.8 Psychology of Software Tester 70
3.9 Test-Driven Development 72
3.10 E-Commerce Testing 72
3.11 Traceability of Requirements 74
3.12 Software Maintenance and Evolution 74
3.13 Software Test Tools 75
3.14 Review Questions 77
3.15 Summary 77
References 78

**4 Static Testing** . . . 79
4.1 Introduction . . . 79
4.2 Economic Benefits of Software Inspections . . . 81
4.3 Informal Reviews . . . 82
4.4 Structured Walk-through . . . 83
4.5 Semi-formal Review Meeting . . . 83
4.6 Fagan Inspections . . . 86
4.6.1 Fagan Inspection Guidelines . . . 88
4.6.2 Inspectors and Roles . . . 89
4.6.3 Inspection Entry Criteria . . . 89
4.6.4 Preparation . . . 90
4.6.5 The Inspection Meeting . . . 90
4.6.6 Inspection Exit Criteria . . . 91
4.6.7 Issue Severity . . . 91
4.6.8 Defect Type . . . 92
4.7 Automated Code Inspections . . . 94
4.8 Review Questions . . . 96
4.9 Summary . . . 96
References . . . 97

**5 Software Test Planning** . . . 99
5.1 Introduction . . . 99
5.2 Test Estimation . . . 102
5.2.1 Estimation Techniques . . . 103
5.2.2 Work Breakdown Structure . . . 103
5.3 Test Planning and Scheduling . . . 103
5.4 Risk Management in Testing . . . 107
5.5 Dedicated Test Plans . . . 109
5.6 Monitoring and Control . . . 110
5.6.1 Managing Issues, Change Requests, and Defects . . . 111
5.7 Project Governance During Testing . . . 112
5.8 Test Reporting . . . 112
5.9 Lessons Learned and Project Closure . . . 114
5.10 Configuration Management . . . 115
5.11 Review Questions . . . 115
5.12 Summary . . . 115
Reference . . . 116

**6 Test Case Analysis and Design** . . . 117
6.1 Introduction . . . 117
6.2 Requirement Engineering . . . 119
6.3 Test Case Design Techniques . . . 120
6.3.1 Black Box Testing . . . 120
6.3.2 White Box Testing . . . 125
6.3.3 Experienced-Based Testing . . . 127

6.4 Test Case Specification . . . 128
6.5 Requirement Traceability . . . 130
6.6 Review Questions . . . 131
6.7 Summary . . . 131
Reference . . . 132

**7 Test Execution and Management** . . . 133
7.1 Introduction . . . 133
7.2 Test Planning . . . 135
7.2.1 Test Team Organization . . . 135
7.3 Test Execution . . . 136
7.4 Managing Defects . . . 137
7.5 Managing Change Requests . . . 139
7.6 Test Monitoring and Control . . . 140
7.7 Risk Management . . . 141
7.8 Test Reporting . . . 141
7.9 Test Completion Criteria . . . 143
7.10 Review Questions . . . 143
7.11 Summary . . . 144

**8 Test Outsourcing** . . . 145
8.1 Introduction . . . 145
8.2 Planning and Requirements . . . 147
8.3 Identifying Suppliers . . . 147
8.4 Prepare and Issue RFP . . . 148
8.5 Evaluate Proposals and Select Supplier . . . 149
8.6 Formal Agreement . . . 149
8.7 Managing the Supplier . . . 150
8.8 Acceptance Testing . . . 151
8.9 Rollout and Customer Support . . . 151
8.10 Review Questions . . . 152
8.11 Summary . . . 152

**9 Test Metrics and Problem-Solving** . . . 153
9.1 Introduction . . . 153
9.2 The Goal, Question, Metric Paradigm . . . 154
9.3 Metrics for Testing . . . 156
9.3.1 Customer Satisfaction Metrics . . . 157
9.3.2 Project Management Metrics for Testing . . . 158
9.3.3 Test Execution Metrics . . . 159
9.3.4 Customer Care Metrics . . . 164
9.3.5 Miscellaneous Metrics . . . 166
9.4 Implementing a Metrics Program . . . 167
9.4.1 Data Gathering for Metrics . . . 168
9.5 Problem-Solving Techniques . . . 169
9.5.1 Fishbone Diagram . . . 171

9.5.2 Histograms . . . 172
9.5.3 Pareto Chart . . . 173
9.5.4 Trend Graphs . . . 175
9.5.5 Scatter Graphs . . . 176
9.5.6 Metrics and Statistical Process Control . . . 176
9.6 Review Questions . . . 177
9.7 Summary . . . 178
References . . . 178

**10 Software Testing Tools** . . . 181
10.1 Introduction . . . 181
10.2 Test Management Tools . . . 184
10.2.1 Estimation and Scheduling Tools . . . 186
10.3 Static Code Analysis Tools . . . 187
10.4 Requirements and Test Design Tools . . . 189
10.5 Test Execution Tools . . . 191
10.5.1 Tools for Regression Testing . . . 193
10.6 Tools for Defect Tracking . . . 194
10.7 Test Performance and Monitoring Tools . . . 195
10.8 Tools for Testing in Agile World . . . 196
10.9 Tools for Configuration Management . . . 197
10.10 Review Questions . . . 197
10.11 Summary . . . 198
Reference . . . 198

**11 Test Process Improvement** . . . 199
11.1 Introduction . . . 199
11.2 Software Process Improvement . . . 200
11.2.1 What Is a Software Process? . . . 202
11.2.2 Benefits of Software Process Improvement . . . 203
11.2.3 Software Process Improvement Models . . . 204
11.2.4 Process Mapping . . . 205
11.2.5 Process Improvement Initiatives . . . 206
11.2.6 Barriers to Success . . . 207
11.2.7 Setting Up an Improvement Initiative . . . 208
11.2.8 Appraisals . . . 208
11.3 Test Process Improvement Models . . . 210
11.3.1 TMM*i* Model . . . 211
11.3.2 TMap Next Model . . . 212
11.3.3 TPI Next Model . . . 214
11.3.4 STEP Model . . . 215
11.3.5 CTP Model . . . 216
11.3.6 PDCA Model . . . 217
11.3.7 CMMI Model . . . 217

11.4 Review Questions . . . . . . . . . . 219
11.5 Summary . . . . . . . . . . 219
References . . . . . . . . . . 220

12 **Testing in the Agile World** . . . . . . . . . . 221
12.1 Introduction . . . . . . . . . . 221
12.2 Scrum Methodology . . . . . . . . . . 225
12.2.1 User Stories . . . . . . . . . . 226
12.2.2 Estimation in Agile . . . . . . . . . . 227
12.2.3 Pair Programming . . . . . . . . . . 228
12.3 Software Testing in Agile . . . . . . . . . . 229
12.3.1 Test-Driven Development . . . . . . . . . . 230
12.3.2 Agile Test Principles . . . . . . . . . . 231
12.4 Review Questions . . . . . . . . . . 232
12.5 Summary . . . . . . . . . . 232
Reference . . . . . . . . . . 233

13 **Verification of Safety-Critical Systems** . . . . . . . . . . 235
13.1 Introduction . . . . . . . . . . 235
13.2 Software Reliability . . . . . . . . . . 237
13.3 Software Dependability . . . . . . . . . . 240
13.4 Formal Methods . . . . . . . . . . 242
13.5 Cleanroom Methodology . . . . . . . . . . 244
13.6 Formal Methods and Testing . . . . . . . . . . 245
13.7 UML and Testing . . . . . . . . . . 246
13.7.1 Model Checking and Testing . . . . . . . . . . 247
13.8 Review Questions . . . . . . . . . . 248
13.9 Summary . . . . . . . . . . 249
References . . . . . . . . . . 250

14 **Legal, Ethical, and Professional Aspects of Testing** . . . . . . . . . . 251
14.1 Introduction . . . . . . . . . . 251
14.2 Business Ethics . . . . . . . . . . 252
14.2.1 What Is Computer Ethics? . . . . . . . . . . 254
14.2.2 The Ethical Software Tester . . . . . . . . . . 255
14.3 Professional Responsibility of Software Engineers and Testers . . . . . . . . . . 256
14.3.1 ACM Code of Professional Conduct and Ethics . . . . . 257
14.4 Legal Aspects of Testing . . . . . . . . . . 257
14.4.1 Legal Impacts of Failure . . . . . . . . . . 259
14.4.2 Lawsuits and Professional Negligence . . . . . . . . . . 259
14.4.3 The Law of Tort and Testing . . . . . . . . . . 260
14.5 Legal Aspects of Test Outsourcing . . . . . . . . . . 261
14.6 Licenses for Test Tools . . . . . . . . . . 263

14.7 Testing and Prevention of Computer Crime . . . . . . . . . . . . . . 264
14.7.1 Testing and Hacking . . . . . . . . . . . . . . . . . . . . . . . 265
14.8 Review Questions . . . . . . . . . . . . . . . . . . . . . . . . . . . . . . 268
14.9 Summary . . . . . . . . . . . . . . . . . . . . . . . . . . . . . . . . . . . 269
References . . . . . . . . . . . . . . . . . . . . . . . . . . . . . . . . . . . . 270

**15 Configuration Management** . . . . . . . . . . . . . . . . . . . . . . . . . 271
15.1 Introduction . . . . . . . . . . . . . . . . . . . . . . . . . . . . . . . . . 271
15.2 Configuration Management System . . . . . . . . . . . . . . . . . . . 275
15.2.1 Identify Configuration Items . . . . . . . . . . . . . . . . . . . 276
15.2.2 Document Control Management . . . . . . . . . . . . . . . . 276
15.2.3 Source Code Control Management . . . . . . . . . . . . . . 277
15.2.4 Configuration Management Plan . . . . . . . . . . . . . . . 277
15.3 Change Control . . . . . . . . . . . . . . . . . . . . . . . . . . . . . . . 278
15.4 Configuration Management Audits . . . . . . . . . . . . . . . . . . . 279
15.5 Configuration Management in Testing . . . . . . . . . . . . . . . . . 280
15.6 Review Questions . . . . . . . . . . . . . . . . . . . . . . . . . . . . . . 281
15.7 Summary . . . . . . . . . . . . . . . . . . . . . . . . . . . . . . . . . . . 282

**16 Epilogue** . . . . . . . . . . . . . . . . . . . . . . . . . . . . . . . . . . . . 283
16.1 The Future of Software Testing . . . . . . . . . . . . . . . . . . . . . 285

**Glossary** . . . . . . . . . . . . . . . . . . . . . . . . . . . . . . . . . . . . . . 287

**Index** . . . . . . . . . . . . . . . . . . . . . . . . . . . . . . . . . . . . . . . 291

# List of Figures


| | | |
|---|---|---|
| Fig. 1.1 | Standish research—project cost estimation accuracy in 1998 | 2 |
| Fig. 1.2 | Shewhart's control chart | 8 |
| Fig. 1.3 | Shewhart's PDCA cycle | 8 |
| Fig. 1.4 | W.E. Deming | 9 |
| Fig. 1.5 | Joseph Juran | 12 |
| Fig. 1.6 | Cost of poor quality—% of sales | 13 |
| Fig. 1.7 | Estimation accuracy—breakthrough and control | 14 |
| Fig. 1.8 | Watts Humphrey. Courtesy of Watts Humphrey | 16 |
| Fig. 1.9 | Cost of quality | 23 |
| Fig. 1.10 | Customer satisfaction process | 25 |
| Fig. 1.11 | Customer satisfaction metrics | 27 |
| Fig. 2.1 | Standish report—results of 1995 and 2009 Survey | 35 |
| Fig. 2.2 | Waterfall V lifecycle model | 41 |
| Fig. 2.3 | SPIRAL lifecycle model. Public domain | 42 |
| Fig. 2.4 | Rational unified process | 43 |
| Fig. 3.1 | Simplified test process | 62 |
| Fig. 3.2 | Number of paths through a trivial program | 69 |
| Fig. 3.3 | Psychology of software tester. Public domain | 71 |
| Fig. 3.4 | Automated testing tools. Creative Commons | 76 |
| Fig. 4.1 | Michael Fagan | 80 |
| Fig. 4.2 | Template for semi-formal review | 85 |
| Fig. 4.3 | Example of an inspection meeting (public domain) | 86 |
| Fig. 4.4 | Defect types in a project (ODC) | 94 |
| Fig. 4.5 | Template for Fagan inspection | 95 |
| Fig. 5.1 | Simple process map for test planning | 105 |
| Fig. 5.2 | Sample Microsoft Project schedule | 106 |
| Fig. 5.3 | Risk categories | 108 |
| Fig. 5.4 | PRINCE2 project board | 113 |
| Fig. 5.5 | Project management triangle | 114 |
| Fig. 6.1 | Test case specification | 120 |
| Fig. 6.2 | State diagram for PIN authentication | 124 |
| Fig. 6.3 | Use-case diagram of ATM | 125 |
| Fig. 7.1 | Organization of a test team | 135 |

Fig. 7.2 Computer bug . . . . . . 138
Fig. 7.3 Test monitoring and control process map . . . . . . 140
Fig. 8.1 Legal agreement . . . . . . 149
Fig. 9.1 GQM example . . . . . . 155
Fig. 9.2 Customer survey arrivals . . . . . . 157
Fig. 9.3 Customer satisfaction measurement . . . . . . 157
Fig. 9.4 Schedule timeliness metric for testing . . . . . . 158
Fig. 9.5 Effort timeliness metric for testing . . . . . . 159
Fig. 9.6 Total number of issues in project . . . . . . 160
Fig. 9.7 Open issues in project . . . . . . 160
Fig. 9.8 Age of open defects in project . . . . . . 161
Fig. 9.9 Problem arrivals per month . . . . . . 161
Fig. 9.10 Phase containment effectiveness . . . . . . 162
Fig. 9.11 Test status . . . . . . 162
Fig. 9.12 Cumulative defects . . . . . . 163
Fig. 9.13 Problem arrival and closure . . . . . . 163
Fig. 9.14 Status of problem . . . . . . 163
Fig. 9.15 Customer queries (arrivals/closures) . . . . . . 164
Fig. 9.16 Outage time per customer . . . . . . 165
Fig. 9.17 Availability of system per month . . . . . . 166
Fig. 9.18 CMMI maturity in current year . . . . . . 166
Fig. 9.19 Fishbone cause-and-effect diagram of high number of defects . . . . . . 171
Fig. 9.20 Histogram . . . . . . 173
Fig. 9.21 Pareto chart outages . . . . . . 174
Fig. 9.22 Trend chart estimation accuracy . . . . . . 175
Fig. 9.23 Scatter graph amount inspected rate/error density . . . . . . 176
Fig. 9.24 Estimation accuracy and control charts . . . . . . 177
Fig. 10.1 HP Quality Center . . . . . . 185
Fig. 10.2 LDRA Code Coverage Analysis Report . . . . . . 188
Fig. 10.3 IBM Rational DOORS Tool . . . . . . 190
Fig. 10.4 Bugzilla: Creative Commons . . . . . . 194
Fig. 10.5 Apache JMeter: Creative Commons . . . . . . 196
Fig. 11.1 Steps in process improvement . . . . . . 201
Fig. 11.2 Process as glue for people, procedures and tools . . . . . . 202
Fig. 11.3 Sample process map . . . . . . 203
Fig. 11.4 ISO 9001 quality management system . . . . . . 205
Fig. 11.5 Continuous improvement cycle . . . . . . 207
Fig. 11.6 Appraisals . . . . . . 209
Fig. 11.7 TMMi maturity levels . . . . . . 212
Fig. 11.8 TMap lifecycle model . . . . . . 213
Fig. 11.9 TPI model . . . . . . 214
Fig. 11.10 Phases of STEP model . . . . . . 215
Fig. 12.1 Agile Dog. Creative Commons . . . . . . 222

Fig. 12.2 Scrum framework. Creative Commons . . . . . . . . . . . . . . . . . . 225
Fig. 12.3 User story map . . . . . . . . . . . . . . . . . . . . . . . . . . . . . . . . . . . . 227
Fig. 12.4 Pair programming. Creative Commons . . . . . . . . . . . . . . . . . . 228
Fig. 13.1 Grafenrheinfeld Nuclear Power Plant. Germany. Creative Commons . . . . . . . . . . . . . . . . . . . . . . . . . . . . . . . . . 236
Fig. 13.2 Formal signing of the treaty of Versailles in 1919. Public Domain . . . . . . . . . . . . . . . . . . . . . . . . . . . . . . . . . . . . 242
Fig. 13.3 Cleanroom in semiconductor manufacturing. Public Domain . . . . . . . . . . . . . . . . . . . . . . . . . . . . . . . . . . . . 245
Fig. 13.4 Deriving tests from abstract model . . . . . . . . . . . . . . . . . . . . . . 246
Fig. 13.5 Model checking . . . . . . . . . . . . . . . . . . . . . . . . . . . . . . . . . . . . 247
Fig. 14.1 Corrupt legislation. 1896. Public domain . . . . . . . . . . . . . . . . . 253
Fig. 14.2 Legal contract. Creative Commons . . . . . . . . . . . . . . . . . . . . . . 262
Fig. 14.3 Hacker at work on backlit keyboard. Creative Commons . . . . . 266
Fig. 15.1 Simple process map for change requests. . . . . . . . . . . . . . . . . . 279
Fig. 15.2 Simple process map for configuration management . . . . . . . . . 280

# List of Tables


Table 1.1 ISO 9126 quality characteristics . . . . . 5
Table 1.2 Shewhart cycle . . . . . 9
Table 1.3 Deming 14-step programme . . . . . 11
Table 1.4 Deming—five deadly diseases . . . . . 12
Table 1.5 Juran's 10-step programme for quality planning . . . . . 13
Table 1.6 Juran's breakthrough and control . . . . . 14
Table 1.7 Crosby's 14-step programme . . . . . 15
Table 1.8 Crosby's maturity grid . . . . . 16
Table 1.9 Cost of quality categories . . . . . 23
Table 1.10 Sample customer satisfaction questionnaire . . . . . 26
Table 1.11 Total quality management . . . . . 28
Table 3.1 Types of testing . . . . . 64
Table 3.2 Test levels . . . . . 64
Table 3.3 Simple test schedule . . . . . 66
Table 4.1 Informal review . . . . . 82
Table 4.2 Structured walk-throughs . . . . . 83
Table 4.3 Activities for semi-formal review meeting . . . . . 84
Table 4.4 Overview Fagan inspection process . . . . . 87
Table 4.5 Strict Fagan inspection guidelines . . . . . 88
Table 4.6 Tailored (relaxed) Fagan inspection guidelines . . . . . 88
Table 4.7 Inspector roles . . . . . 89
Table 4.8 Fagan entry criteria . . . . . 90
Table 4.9 Inspection meeting . . . . . 91
Table 4.10 Fagan exit criteria . . . . . 92
Table 4.11 Issue severity . . . . . 92
Table 4.12 Classification of defects in Fagan inspections . . . . . 93
Table 4.13 Classification of ODC defect types . . . . . 93
Table 5.1 Estimation techniques . . . . . 104
Table 5.2 Example of work breakdown structure for test estimation . . . . . 105
Table 5.3 Sample test planning checklist . . . . . 107
Table 5.4 Risk management activities . . . . . 109
Table 5.5 Project board roles and responsibilities . . . . . 113
Table 6.1 Test design techniques . . . . . 121
Table 6.2 Decision table with business rules . . . . . 123

Table 6.3 Planning section in dedicated test plan.................... 128
Table 6.4 Template for test case .................................. 129
Table 6.5 Sample trace matrix ..................................... 130
Table 7.1 Test management activities .............................. 134
Table 7.2 Activities in managing change requests .................. 139
Table 7.3 Test status for project.................................. 142
Table 7.4 Quality status for project .............................. 142
Table 7.5 Key risks for project key risks.......................... 142
Table 8.1 Supplier selection and management ....................... 146
Table 9.1 Implementing metrics..................................... 168
Table 9.2 Goals and questions...................................... 168
Table 9.3 Phase containment effectiveness.......................... 168
Table 10.1 Advantages of test tools................................ 182
Table 10.2 Tool evaluation table .................................. 183
Table 10.3 Types of tools for testing.............................. 183
Table 10.4 Tools for requirements development and management....... 190
Table 11.1 Test process improvement models ........................ 210
Table 11.2 TMM*i* model ........................................... 211
Table 11.3 CMMI requirements for verification...................... 218
Table 11.4 CMMI requirements for validation........................ 218
Table 12.1 Agile test principles .................................. 232
Table 13.1 Software reliability testing ........................... 239
Table 13.2 Dimensions of dependability ............................ 240
Table 13.3 Model-checking process.................................. 248
Table 14.1 Ten commandments on computer ethics..................... 255
Table 14.2 Professional responsibilities of software engineers
and testers ........................................................ 258
Table 14.3 ACM code of conduct (general obligations)............... 258
Table 14.4 Types of lawsuits....................................... 260
Table 15.1 Features of good configuration management .............. 272
Table 15.2 Symptoms of poor configuration management............... 273
Table 15.3 Software configuration management activities ........... 274
Table 15.4 Build plan for project.................................. 274
Table 15.5 CMMI requirements for configuration management.......... 275
Table 15.6 Sample configuration management audit checklist......... 281

# 1 Fundamentals of Software Quality

**Key Topics**

Shewhart
Deming
Juran
Crosby
Watts Humphrey
Metrics
Problem-solving
Cost of quality
Process improvement
Customer satisfaction

## 1.1 Introduction

The mission of a software company is to develop high-quality innovative products and services at a competitive price to its customers and to do so ahead of its competitors. This requires a clear vision of the business, a culture of innovation, an emphasis on quality, detailed knowledge of the business domain, and a sound product development strategy.

It requires a focus on software quality and customer satisfaction, and quality must be built into the software product so that customers remain loyal to the company. Customers have very high expectations on quality and expect high-quality software products to be consistently delivered on time and on budget.


G. O'Regan, *Concise Guide to Software Testing*, Undergraduate Topics in Computer Science,
https://doi.org/10.1007/978-3-030-28494-7_1

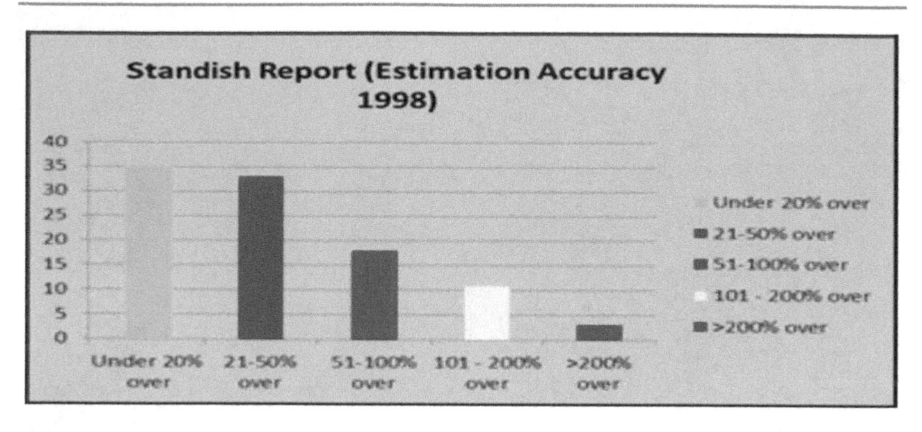


**Fig. 1.1** Standish research—project cost estimation accuracy in 1998

The focus on quality requires effective software processes to be in place so that quality software may be consistently produced.

Software testing plays a key role both in building quality into the software and in verifying that the desired quality has been achieved. Quality improvement is essential, and a focus on industrial best practice and emerging technologies assists in performance improvement.

The history of quality and some of the key people who have contributed to the quality movement are discussed later in the chapter. This includes well-known quality gurus such as Shewhart, Deming, Juran, and Crosby. These figures played an important role in promoting quality and transforming struggling manufacturing companies. Watts Humphrey is considered the father of software quality, and his important contributions to software process improvement are discussed.

The Standish Group Research (1999) (Fig. 1.1) on project cost overruns in the US during 1998 indicate that 33% of projects were between 21 and 50% over estimate, 18% were between 51 and 100% over estimate, and 11% of projects were between 101 and 200% overestimate.[1]

Projects sometimes fail, and there are many examples of projects being abandoned prior to completion. For example, the Taurus project at the London stock exchange is a well-known disaster. The project was eventually abandoned, and at that stage, it was 11 years late and had cost the city of London hundreds of millions of pounds (Manley 1995).

It is essential that requirements are properly managed as uncontrolled changes to requirements may have a negative effect on the project. It may be necessary to accept a late change to the requirements, but there are corresponding risks to the project schedule and quality. However, a good requirements process will ensure that changes to the requirements are minimized and controlled, and the

[1]The study was from the mid/late 1990 and recent reports from the Standish Group show good improvement trends.

requirements process will often include prototyping or joint user reviews to ensure that the requirements are actually those desired by the customer.

The implementation of the requirements involves design, development, and testing activities. It may also involve the production of user manuals and training materials as well as the technical documentation. Quality must be built into the software, and the goal of the testing activities is to verify the correctness of the software. The project manager is responsible for delivering the project on time and for recovering the schedule when it falls behind.

Engineers have constructed bridges for several millennia, and bridge building is considered a mature engineering activity. However, occasionally civil engineering projects fall behind schedule or suffer design flaws. For example, the infamous Tacoma Narrows Bridge (or Galloping Gertie as it was known) collapsed in 1940 due to a design flaw.

The Tacoma Narrows Bridge was known for its tendency to sway in windstorms. The shape of the bridge was like that of an aircraft wing, and under windy conditions, it would generate sufficient lift to become unstable. A large windstorm in November 1940 caused catastrophic failure. The significance of the Tacoma Bridge is its collapse and the subsequent investigation by engineers. They realized that aerodynamical forces in suspension bridges were not sufficiently understood in the design of the bridge and that new research was needed. It was recommended that wind tunnel tests be used to aid in the design of the replacement bridge.

Software engineering is a less mature field than civil engineering, and it is only in more recent times that investigations and recommendations from software projects have become part of the software development process. The study of software engineering has led to new theories and understanding of software development.

## 1.2 History of Software Failures

There are many examples of software failures in the literature. These include the year 2000 (or Y2K) problem which was a design flaw in the representation of the date with two digits; the Intel Pentium microprocessor bug which referred to a floating-point problem on an Intel microprocessor back in 1994; the Ariane 5 launcher disaster was due to an operand error that resulted from the conversion of a 64-bit floating-point number to a 16-bit signed integer number. Software failures may cause major problems and adversely affect the customer's business. They may lead to credibility issues and damage to the customer relationship.

The Y2K bug is historical and part of computer science folklore. The event on 1 January 2000 had minimal impact on the world economy. However, organizations spent large sums of money in identifying all code with a year 2000 impact; changing the representation of the date from 2 digits to 4 digits; and verifying the correctness of the changes made. The worldwide cost of this was in billions of dollars.

The Intel response to a famous microprocessor bug back in 1994 inflicted temporary damage on the reputation of the company. Intel was slow to acknowledge the floating-point problem and in providing adequate information. This led to damage in its reputation and hundreds of millions of dollars to replace the flawed microprocessors.

The Ariane 5 led to major embarrassment and damage to the credibility of the European Space Agency (ESA). The maiden flight of the Ariane 5 launcher ended in failure on 4 June 1996, after a flight time of 40 s. The first 37 s of flight proceeded as normal. However, the launcher then veered off its flight path, broke up, and exploded. An independent inquiry board investigated the cause of the failure, and the report and recommendations to prevent a future failure are described in Lions (1996).

The inquiry noted that the failure of the inertial reference system was followed immediately by a failure of the backup inertial reference system. The problem was traced to a software failure due to an operand error resulting from the conversion of a 64-bit floating-point number to a 16-bit signed integer value number. The floating-point number was too large to be represented in the 16-bit number, and this resulted in the operand error.

The inertial reference system and the backup reference system reported failure due to the software exception. The operand error occurred owing to an exceptionally high value related to the horizontal velocity, and this was due to the fact that the early part of the trajectory of the Ariane 5 differed from the earlier Ariane 4, and required a higher horizontal velocity. The inquiry board made a series of recommendations to prevent a reoccurrence of similar problems.

These failures indicate that software quality needs to be a key driving force in any organization. The effect of software failure may result in huge costs to correct the software (e.g. Y2K), negative perception of a company and large replacement costs (e.g. Intel microprocessor problem), or the loss of a valuable communications satellite and all the costs associated with this (e.g. Ariane 5).

## 1.3 Background to Software Quality

Customers today have very high-quality and reliability expectations and expect companies to adhere to very high standards. There are many quality software products in the marketplace; however, the task of consistently producing high-quality software products is non-trivial. Even the most respected organizations occasionally deliver software that contains defects, or ship products late due to quality problems. Defects may cause minor irritation to a customer, loss of credibility, or lead to injury or loss of life.

The late delivery of a product leads to extra costs, and it may adversely affect the customer's revenue, profitability, and business planning. Consequently, it is essential to have a robust process to consistently develop high-quality software on

**Table 1.1** ISO 9126 quality characteristics

| Characteristic | Description |
|---|---|
| Functionality | This indicates the extent to which the required functionality is available in the software |
| Reliability | This indicates the extent to which the software is reliable |
| Usability | This indicates the extent to which the users of the software judge it to be easy to use |
| Efficiency | This characteristic indicates the efficiency of the software |
| Maintainability | This indicates the extent to which the software product is easy to modify and maintain |
| Portability | This indicates the ease of transferring the software to a different environment |

time and within budget. The influential papers by Fred Brooks in Brooks (1975, 1986) suggest that there is no silver bullet to do this, and that instead, the focus needs to be on incremental improvement to processes and tools.

### 1.3.1 What Is Software Quality?

There are various definitions of quality such as the definition proposed by Philip Crosby as "*conformance to the requirements*". This definition does not take the intrinsic difference in quality of products into account in judging the quality of the product. For example, this definition might suggest that a *Mercedes* car is of the same quality as a *Lada* car.[2] Further, the definition does not consider whether the requirements are actually appropriate for the product.

Juran defines quality as "*fitness for use*", and this is a better definition, although it does not provide a mechanism to judge better quality when two products are equally fit to be used. The ISO 9126 standard for information technology (ISO/IEC 1991) is a framework for the evaluation of software product quality. It defines six product quality characteristics (Table 1.1), which indicate the extent to which a software product may be judged to be of a high quality by the customers.

### 1.3.2 Early Quality Management

In the Middle Ages, a craftsman was responsible for the complete development of a product from its conception to delivery to the customer. This led to a strong sense of pride and ownership of the quality of the product, and apprentices joined craftsmen to learn the skills of the trade.

The Industrial Revolution led to a change to this traditional paradigm, and labour became highly organized with workers responsible for a particular part of the

[2]Most rational people would judge the Mercedes to be of superior quality.

manufacturing process. The sense of ownership and the pride of workmanship in the product were diluted, as workers were now responsible only for their portion of the product, and not the quality of the product as a whole.

This led to a requirement for more stringent management practices, including planning, organizing, implementation, and control. It inevitably led to a hierarchy of labour with various functions identified and a reporting structure for the various functions. Supervisor controls were needed to ensure that quality and productivity issues were addressed.

### 1.3.3 Total Quality Management

Total quality management (TQM) is a modern approach to quality management, and this management philosophy involves customer focus, process improvement, developing a culture of quality within the organization and developing a measurement and analysis program. It emphasizes that customers have rights and quality expectations, which should be satisfied, and that everyone in the organization is both a customer and has customers.

It is a *holistic approach* and requires that all functions, in the organization, follow high standards. Quality needs to be built into the product by ensuring that quality is addressed at every step in the process.

It requires total commitment from the top management, and that all staff be trained in quality management and participate in quality improvement. It requires that a commitment to quality be instilled in all staff, and that the focus within the organization changes from *firefighting* to *fire prevention*. Problem-solving is used to identify the root causes of problems, and corrective action is taken to prevent their re-occurrence.

### 1.3.4 Software Quality Control

Software quality control is concerned with activities to ensure that the end product satisfies the functional and non-functional requirements and is fit for purpose. It includes inspections and testing to verify that the deliverables produced satisfy their requirements. Inspections typically consist of a formal review of a deliverable by independent experts, and the objective is to identify defects within the work product and to provide confidence in its correctness.

Inspections in a manufacturing environment are quite different in that they take place at the end of the production cycle and do not offer a mechanism to build quality into the product. Instead, the defective products are removed from the batch and reworked. There is a growing trend towards quality sampling at the early phases of a manufacturing process to minimize reworking of defective products.

Software testing consists of "*white box*" or "*black box*" testing techniques, and the testing activities include *unit*, *system*, *performance*, and *acceptance testing*. The testing is quite methodical and includes a comprehensive set of manual or automated test cases. The *verification* and *validation* activities involve the execution of the defined tests and the correction of any failed or blocked tests.

The cost of correction of a defect is related to the phase in which it is detected in the lifecycle. Errors detected in phase are the least expensive to correct, and defects detected out of phase become increasingly expensive to correct. The most expensive defect is that of a requirements' defect identified by the customer, as its correction may involve changes to the requirements, design, and code. Testing will be required as well as a fix release for the customer. There is further overhead in project management, configuration management, and in communication with the customer.

It is, therefore, highly desirable to capture defects as early as possible in the software lifecycle to minimize the effort required to correct. Modern software engineering places emphasis on defect prevention and in learning lessons from the defects. This approach is adopted from manufacturing environments and consists of formal causal analysis meetings to brainstorm and identify root causes of problems and to define the corrective actions necessary to prevent reoccurrence. The actions are then implemented and tracked to completion.

## 1.4 History of Quality

This section considers the ideas of several pioneers who have influenced the quality field. These include Walter Shewhart, W. Edwards Deming, Joseph Juran, and Philip Crosby. We also discuss the influence of Watts Humphrey who is considered the father of software quality.

### 1.4.1 Shewhart

Walter Shewhart was Statistician at AT&T Bell Laboratories (or Western Electric Co. as it was known in the 1920s). He is regarded as Founder of statistical process control (SPC), which remains important today in monitoring and controlling a process (Fig. 1.2). Shewhart developed a control chart, which is used to control the process, with upper and lower limits for process performance specified. The process is under control if it is performing within these limits.

Shewhart's ideas were applied to the Capability Maturity Model (CMM) in the late 1980s as a way to control key software processes. Statistical process control (SPC) plays an important role in process improvement and in ensuring that process performance is acceptable. It is used to minimize variability in process performance, as variability in the process affects product quality. SPC involves the analysis of control charts so that the cause of variability can be identified and eliminated. Deming and Juran worked with Shewhart at Bell Labs in the 1920s.

The Shewhart model is a systematic approach to problem-solving and process control. It consists of four steps that are used for continuous process improvement,

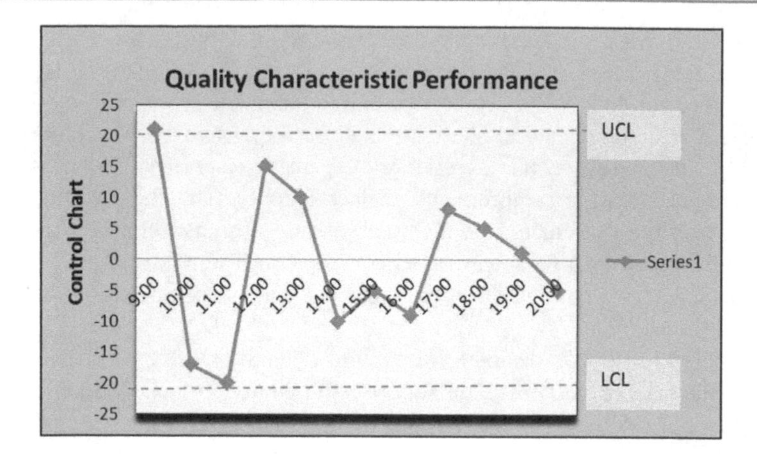


**Fig. 1.2** Shewhart's control chart

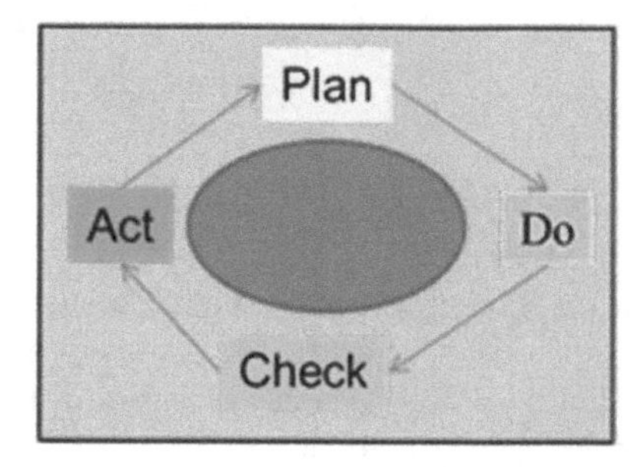


**Fig. 1.3** Shewhart's PDCA cycle

which are *plan, do, check, act* (Fig. 1.3). It is known as the *"PDCA model"* or Shewhart's model and is described in Table 1.2.

Shewhart argued that quality and productivity improve as process variability is reduced. His influential book, *The Economic control of quality of manufactured product* (Shewhart 1931), outlines the methods of statistical process control to reduce process variability. It predicted that productivity would improve as process variability was reduced, and this was verified by Japanese engineers in the 1950s.

This led to productivity improvements and increased market share for Japanese companies. Today, companies around the world recognize the importance of placing quality at the heart of the organization.

### 1.4.2 Deming

W. Edwards Deming (Fig. 1.4) was a major figure in the quality movement. He was influenced by Shewhart's work on statistical process control, and Deming's approach was adopted in post-Second World War Japan. He played an important role in transforming Japan industry.

**Table 1.2** Shewhart cycle

| Step | Description |
|---|---|
| Plan | This step identifies an improvement opportunity and outlines the problem or process that will be addressed<br>– Select the problem to be solved<br>– Describe current process<br>– Identify the possible causes of the problem<br>– Find the root cause<br>– Develop an action plan to correct the root cause |
| Do | This step involves carrying out the improvement actions, and it may involve a pilot of the proposed changes to the process |
| Check | This step involves checking the results obtained to determine their effectiveness |
| Act | This step includes the analysis of the results to adjust process performance to achieve the desired results |

**Fig. 1.4** W.E. Deming

Deming argued that it is not sufficient for everyone in the organization to be doing one's best: instead, what is required is that there be a consistent purpose and direction in the organization. That is, it is first necessary that people know what to do, and there must be a *constancy of purpose* from all individuals to ensure success.

He argued that there is a very strong case for improving quality, as costs will decrease due to less rework, and productivity will increase as less time is spent in reworking defective products. This will enable the company to increase its market share, with better quality and lower prices, and to stay in business. Conversely, companies that fail to address quality issues will lose market share and go out of business. Deming was highly critical of the then American approach to quality and the lack of vision of American management in quality management.

Deming's influential book *Out of the Crisis* (Deming 1986) proposed 14 principles to transform the western style of management of an organization to a quality- and customer-focused organization. These include:

- Constancy of purpose
- Quality built into the product
- Continuous improvement culture.

Deming's ideas are described in more detail in Table 1.3.

Deming argued that there are several diseases that afflict companies in the western world that prevent them for achieving high-quality results. The *"five deadly diseases"* noted by Deming include (Table 1.4).

**Comment (Deming)**

*Deming's programme has been quite influential and has many sound points. His views on slogans in the workplace are in direct opposition to the use of slogans like Crosby's "zero defects". The key point for Deming is that a slogan has no value unless there is a clear method to attain the particular goal described by the slogan.*

### 1.4.3 Juran

Joseph Juran (Fig. 1.5) was a major figure in the quality movement, and he argued for a top-down approach to quality. He defined quality as *"fitness for use"* and argued that quality issues are the direct responsibility of management. Management must ensure that quality is planned, controlled, and improved.

The trilogy of *quality planning*, *control*, and *improvement* is known as the *"Juran Trilogy"* and is usually described by a diagram with time on the horizontal axis and the cost of poor quality on the vertical axis (Fig. 1.6).

Quality planning consists of setting quality goals, developing plans, and identifying the resources needed to achieve the goals. Quality control consists of evaluating performance, setting new goals, and taking appropriate action. Quality improvement consists of improving delivery, eliminating wastage, and improving customer satisfaction. Juran's 10-step programme for quality planning is defined in Juran (1951) and is summarized in Table 1.5.

Juran defined an approach to achieve a new quality performance level that is termed *"Breakthrough and Control"*. It is described pictorially by a control chart showing the old performance level with occasional spikes or random events; what is needed is a breakthrough to a new and more consistent quality performance, i.e. a new performance level with the performance achieved at that level.

The example in Fig. 1.7 presents a breakthrough in developing a more accurate estimation process. Initially, the variation in estimation accuracy is quite large, but as an improved estimation process is put in place, the control limits are narrowed and more consistent estimation accuracy is achieved.

The breakthrough is achieved by a sustained and coordinated effort, and the old performance standard becomes obsolete. The difference between the old and the new performance level is known as the *"chronic disease"* which must be diagnosed and cured. His approach to breakthrough and control is described in Table 1.6.

**Table 1.3** Deming 14-step programme

| Step | Description |
|---|---|
| Constancy of purpose | Companies face short-term and long-term problems. The problems of tomorrow require long-term planning on new products, training, and innovation. This requires R&D and continuous improvement of existing products and services |
| Adopt new philosophy | Deming outlined the *five deadly diseases* that afflicted US companies. These included lack of purpose and an excessive interest in short-term profits |
| Build quality in | Deming argued that performing mass inspections is equivalent to planning for defects, as they are too late to improve quality. Consequently, it is necessary to improve the production process to *build the quality into* the product |
| Price and quality | Deming argued against awarding business on price alone, as the price is meaningless unless there is an objective measure of the quality of the product being purchased |
| Continuous improvement | There must be *continuous improvement in all areas*, including understanding customer requirements, design, manufacturing, and test methods |
| Institute training | The organization must be a learning organization with a training programme to educate management and staff about the company, customer needs, and pride of workmanship in the products. Supervisors and managers need training on the 14-point program |
| Institute leadership | Deming argues that *management is about leadership and not supervision*. Management should work to remove barriers, know the work domain, and seek innovative solutions to resolve quality and other relevant issues |
| Eliminate fear | The presence of fear is a barrier to an open discussion of problems and the identification of solutions or changes to prevent problems from arising |
| Eliminate barriers | The objective here is to break down barriers between different departments and groups. It is not enough for each group to optimize its own area: instead, what is required is for the organization to be working as one team |
| Eliminate slogans | Deming argued that slogans do not help anyone to do a better job. Slogans may potentially alienate staff or encourage cynicism. Deming criticized slogans such as "*Zero Defects*" or "*Do it right the first time*" as inappropriate, as how can it be made right first time if the production machine is defective. Most problems are due to the system rather than the person |
| Eliminate numerical quotas | Deming argued that quotas act as an impediment to improvement in quality, as quotas are normally based on what may be achieved by the average worker. People below the average cannot make the rate, and the result is dissatisfaction and turnover. Thus, there is a fundamental conflict between quotas and pride of workmanship |
| Pride of work | The intention here is to remove barriers that rob people of pride of workmanship (e.g. machines out of order) |

(continued)

**Table 1.3** (continued)

| Step | Description |
|---|---|
| Self improvement | This involves encouraging education and self-improvement for everyone in the company |
| Take action | This requires that management agree on direction using the 14 principles, communicate the reasons for changes to the staff, and train the staff on the 14 principles |

**Table 1.4** Deming—five deadly diseases

| Disease | Description |
|---|---|
| Lack of constancy of purpose | Management is too focused on short-term thinking rather than long-term improvements |
| Emphasis on short-term profit | A company should aim to become the world's most efficient provider of product/service. Profits will then follow |
| Evaluation of performance | Deming is against annual performance appraisal and rating |
| Mobility of management | Mobility of management frequently has a negative impact on quality |
| Excessive measurement | Excessive management by measurement |

**Fig. 1.5** Joseph Juran

### 1.4.4 Crosby

Philip Crosby was a key figure in the quality movement, and his quality improvement grid later influenced the design of the Capability Maturity Model (CMM), which was developed by the Software Engineering Institute. His influential book *Quality is Free* (Crosby 1979) outlines his philosophy of *doing things right the first time*, i.e. the *zero defects* (ZD) program. Quality is defined as *"conformance to the requirements"*, and he argues that people have been conditioned to believe that error is inevitable.

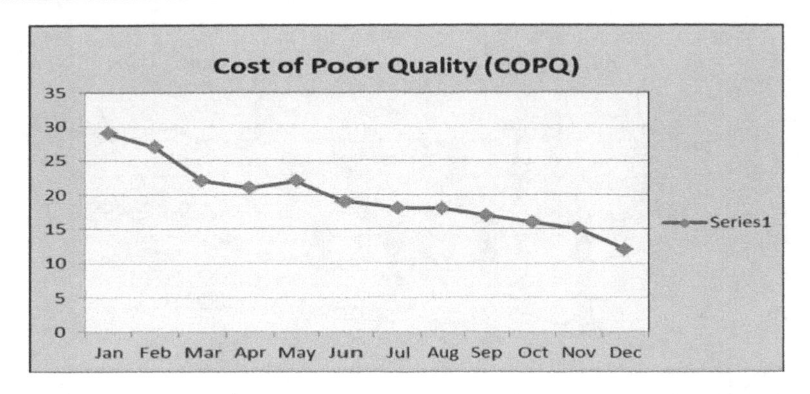


**Fig. 1.6** Cost of poor quality—% of sales

**Table 1.5** Juran's 10-step programme for quality planning

| Step | Description |
|---|---|
| Identify customers | This includes the internal and external customers of an organization; e.g., the testing group is an internal customer, whereas the end-user of the software is an external customer |
| Determine customer needs | Customer needs are generally expressed in the language of the customer's organization. There is a need to elicit and determine the actual desired requirements from discussion and communication with the customer |
| Translate | This involves translating the customer needs into the language of the supplier |
| Units of measurement | This involves defining the measurement units to be used |
| Measurement programme | This involves setting up a measurement programme in the organization, and it includes internal and external measurements of quality and process performance |
| Develop product | This step determines the product features to meet the needs of the customer |
| Optimize product design | The intention is to optimize the design of the product to meet the needs of the customer and supplier |
| Develop process | This involves developing processes that can produce the products to satisfy the customer's needs |
| Optimize process capability | This involves optimizing the capability of the process to ensure that high-quality products are produced |
| Transfer | This involves transferring the process to normal product development operations |

Crosby argued that people in their personal lives do not accept this: for example, it would not be acceptable for nurses to drop a certain percentage of newly born babies. He further argues that the term *"acceptable quality level"* (AQL) is a commitment to produce imperfect material. Crosby notes that defects are due to two main reasons: *lack of knowledge* or a *lack of attention of the individual.*

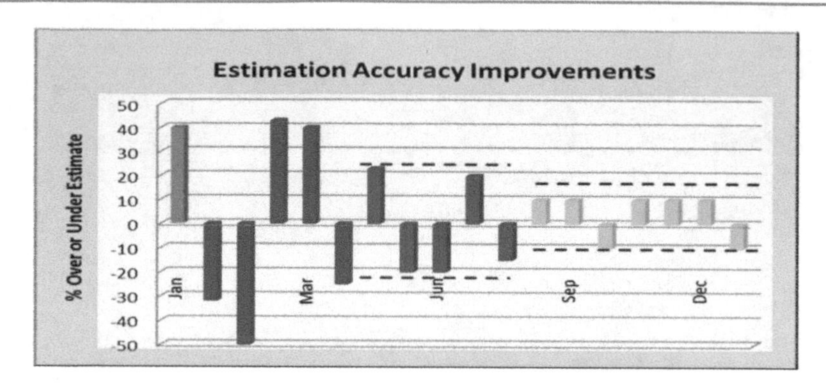


**Fig. 1.7** Estimation accuracy—breakthrough and control

**Table 1.6** Juran's breakthrough and control

| Step | Description |
|---|---|
| Breakthrough in attitude | This involves developing a favourable attitude to quality improvement |
| Pareto | This involves identifying the key areas affecting quality |
| Organization | This involves analysing the problem and coordinating a solution |
| Control | This is concerned with achieving performance at the new level |
| Repeat | This leads to continuous improvement with new performance levels set, and new breakthroughs made to achieve higher performance levels |

He argued that lack of knowledge can be measured and addressed by training, but that lack of attention is a mindset that requires a change of attitude by the individual. The net effect of a successful implementation of a zero defects programme is higher productivity due to less reworking of defective products. Thus, *quality*, in effect, *is free*.

Crosby's approach to achieving the desired quality level of zero defects was to put a quality improvement programme in place. He outlined a 14-step quality improvement programme (Table 1.7). It requires management commitment to be successful, and an organization-wide quality improvement team needs to be set up. A measurement programme is put in place to determine the status and cost of quality within the organization. The cost of quality is then shared with the staff, and corrective actions are identified and implemented. The zero defects programme is communicated to the staff, and one day every year is made a *zero defects day* and is used to emphasize the importance of zero defects to the organization.

Crosby's Quality Management Maturity Grid (Table 1.8) measures the maturity of the current quality system with respect to several quality management categories and highlights areas that require improvement. Six categories of quality management are considered: *management understanding and attitude towards quality, quality organization status, problem handling, the cost of quality, quality improvement actions, and summation of company quality posture.*

**Table 1.7** Crosby's 14-step programme

| Step | Description |
|---|---|
| Management commitment | Management commitment and participation are essential to the success of the quality improvement program. The profile of quality is raised within the organization |
| Quality improvement team | This involves the formation of an organization-wide cross-functional team consisting of representatives from each of the departments |
| Quality measurement | The objective is to determine the status of quality in each area of the company to identify areas where improvements are required |
| Cost of quality evaluation | The cost of quality indicates the financial cost of quality to the organization. It is initially high, but reduces as the quality improvement programme becomes effective |
| Quality awareness | This involves sharing the cost of poor quality with staff and motivating staff to identify corrective actions to deal with quality issues |
| Corrective action | This involves resolving any problems that have been identified and bringing any problems that cannot be resolved to the attention of management |
| Zero defects program | The key point is that zero defects is not a motivation program: instead, it means doing things right the first time, i.e. zero defects |
| Supervisor training | This requires that all supervisors and managers receive training on the 14-step quality improvement program |
| Zero defects day | This involves setting aside one day each year to high-light zero defects and its importance to the company |
| Goal setting | This phase involves getting people to think in terms of goals and how the goals may be achieved |
| Error cause removal | This involves removing any roadblocks or problems that prevent employees from performing error-free work |
| Recognition | This involves recognizing employees who make outstanding contributions to quality improvement |
| Quality councils | This involves bringing quality professionals together on a regular basis to share ideas on quality |
| Do it over again | The principle of continuous improvement is a key part of the programme, as improvement is continuous |

Each category is rated on a 1-to-5 maturity scale which indicates the maturity of the particular category. Crosby's maturity grid was later adapted and applied to the CMM. The five maturity levels of Crosby's grid are:

**Comment (Crosby)**

*Crosby's programme has been quite influential, and his maturity grid has been applied to the software CMM. The ZD part of the programme is difficult to apply to the complex world of software development, where the complexities of the systems to be developed are often the cause of defects rather than the mindset of software professionals (who are generally professional and dedicated to quality). Slogans may be dangerous and potentially unsuitable to some cultures, and a zero defects day may potentially have the effect of de-motivating staff.*

**Table 1.8** Crosby's maturity grid

| Level | Name | Description |
|---|---|---|
| 1. | Uncertainty | Management has no understanding of quality and is likely to blame quality problems on the quality department. Firefighting is prevalent, and problems are fought as they occur. Root causes of problems are not investigated, and there are few organized quality improvement activities |
| 2. | Awakening | Management is beginning to recognize that quality management may be of value, but is unwilling to devote time and money to it. Instead, the emphasis is on appraisal rather than prevention. Teams are set up to address major problems, but long-term solutions are rarely sought |
| 3. | Enlightenment | Management is learning more about quality and is becoming more supportive of quality improvement. The quality department reports to senior management, and implementation of the 14-step quality improvement programme is underway. There is a culture of openness where problems are faced openly and resolved in an orderly way |
| 4. | Wisdom | Management is fully participating in the program and fully understands the importance of quality management. All functions within the organization are open to suggestions for improvement, and problems are identified earlier. Defect prevention is now part of the culture |
| 5. | Certainty | The whole organization is involved in continuous improvement |

## 1.4.5 Watts Humphrey

Watts Humphrey was an American software engineer and vice-president of technical development at IBM. He made important contributions to the software engineering field and is considered the *father of software quality*. He dedicated much of his career to addressing the problems of software development including schedule delays, cost overruns, software quality, and productivity (Fig. 1.8).

He was born in Michigan in 1927 and served in the US Navy and completed a bachelor's degree in physics at the University of Chicago in 1949. He obtained a

**Fig. 1.8** Watts Humphrey. Courtesy of Watts Humphrey

master's degree in physics from the Illinois Institute of Technology (IIT) and an MBA from the University of Chicago.

He took a position with Sylvania in Boston in the early 1950s, and he became Manager of the circuit design group in the company. He recognized the importance of planning and management early in his career, and he joined IBM in 1959 initially as Hardware Architect, but most of his IBM career was in management. He was eventually to become Vice-President of technical development, where he oversaw 4,000 engineers in 15 development centres in over 7 countries. Others at IBM influenced him including Fred Brooks who was Project Manager of the IBM 360 project; Michael Fagan who developed the Fagan inspection methodology; and Harlan Mills who developed the Cleanroom methodology. Humphrey ran the software quality and process group at IBM towards the end of his IBM career and became very interested in software quality.

He retired from IBM in 1986 and joined the newly formed SEI at Carnegie Mellon University. He made a commitment to change the software engineering world by developing sound management principles for the software industry. The SEI has largely fulfilled this commitment, and it has played an important role in enhancing the capability of software organizations throughout the world.

The SEI had a contract from the Department of Defence (DOD) to provide guidance to the military in the selection of capable software subcontractors. This evolved into the book "Managing the Software Process" (Humphry 1989) which describes technical and managerial topics essential for good software engineering. The book was influenced by the ideas of Deming and Juran in statistical process control.

Humphrey established the software process programme at the SEI, and this led to the development of the software Capability Maturity Model (CMM) and its successors. Humphrey asked questions such as:

- How good is the current software process?
- What must I do to improve it?
- Where do I start?

The CMM is a framework to help an organization to understand its current process maturity and to prioritize improvements. The SEI introduced software process assessment and software capability evaluation methods, and these include CBA/IPI and CBA/SCE. The CMM and the associated assessment methods were widely adopted by organizations around the world, and their successors are the CMMI Model and the SCAMPI appraisal methodology.

Humphrey focused his later efforts to developing the Personal Software Process (PSP) and the Team Software Process (TSP). These are approaches that teach engineers the skills they need to make and track plans and to produce high-quality software with zero defects. The PSP helps the individual engineer to collect relevant data for statistical process control, whereas the TSP focuses on teams, and the goal is to assist teams to understand and improve their current productivity and quality of their work.

He received many awards for his contributions to the computing field. He was named the first SEI fellow in 1995 in recognition of his outstanding contribution to the software quality field. He received the 2003 National Medal in Technology and Innovation from President George Bush, and he was named an ACM fellow in 2009 for his outstanding contributions to computing and information technology. He was the author of twelve books in the software engineering field, and he died in 2010.

### 1.4.6 Miscellaneous Quality Gurus

There are several other pioneers in the quality field including *Shingo* who developed his own version of zero defects termed *"Poka-yoke"* (or *defects* = 0). This involves identifying potential error sources in the process and monitoring these for errors. Causal analysis is performed on any errors found, and the root causes are eliminated. This approach leads to the elimination of all errors likely to occur, and thus only exceptional errors should occur. These exceptional errors and their causes are then eliminated. The failure mode and effects analysis (FMEA) methodology is a variant of this. Potential failures to the system or subsystem are identified and analysed, and the causes and effects and probability of failure documented.

*Genichi Taguchi's* definition of quality is quite different. Quality is defined as *"the loss a product causes to society after being shipped, other than losses caused by its intrinsic function"*. Taguchi defines a *loss function* as a measure of the cost of quality; $L(x) = c(x - T)^2 + k$. Taguchi also developed a method for determining the optimum value of process variables which will minimize the variation in a process while keeping a process mean on target.

*Kaoru Ishikawa* did work on *quality control circles* (QCCs). A quality control circle is a small group of employees who do similar work and meet regularly to identify and analyse work-related problems. This involves brainstorming, recommending, and implementing solutions. The problem-solving tools employed include *Pareto analysis*, *fishbone diagrams*, *histograms*, *scatter diagrams*, and *control charts*. A facilitator will train the quality circle team leaders, and the activities in a quality circle include:

- Select problem
- State and restate problem
- Collect facts
- Brainstorm
- Build on each other's ideas
- Choose course of action
- Presentation.

*Armand Feigenbaum* did work in *total quality control* which concerns quality assurance applied to all functions in the organization. It is distinct from total quality management: total quality control is concerned with controlling quality throughout, whereas TQM embodies a philosophy of quality management and improvement involving all staff and functions throughout the organization.

## 1.5 Modern Software Quality Management

The development of high-quality software requires a good software development process to be in place, and this includes best practices in software engineering for:

- Project management
- Estimation
- Risk management
- Requirements' development and management
- Design and development
- Software development lifecycles
- Quality assurance/management
- Software inspections
- Software testing
- Supplier selection and management
- Configuration management
- Customer satisfaction process
- Continuous improvement.

The cost of correction of a defect increases the later that it is detected in the lifecycle. Consequently, it is desirable to detect an error as early as possible and preferably within the phase in which it was created. Software inspections play a key role in detecting defects in-phase, and they are discussed in the next section.

### 1.5.1 Software Inspections

The Fagan inspection process was developed by Michael Fagan of IBM (Fagan 1976), and it aims to identify and remove errors in work products. The process mandates that requirement documents, design documents, source code, and test plans all be formally inspected by experts independent of the author of the deliverable.

There are various *roles* defined in the process including the *moderator* who chairs the inspection. The moderator ensures that all of the inspectors are trained and receive the appropriate materials for the inspection. He/she ensures that sufficient preparation is done, and that the speed of the inspection does not exceed the recommended guidelines. The *reader* reads or paraphrases the particular deliverable; the *author* is the creator of the deliverable and has a special interest in ensuring that it is correct. The *tester* role is concerned with the test viewpoint.

The inspection process will consider whether the design is correct with respect to the requirements, and whether the source code is correct with respect to the design. The errors identified are classified into various types, and the data is generally recorded to enable analysis to be performed on the most common types of errors to yield actions to minimize the re-occurrence of the most common defect types. Software inspections are described in more detail in Chap. 4.

### 1.5.2 Software Testing

Software testing plays a key role in verifying that the software is fit for purpose, and two key types of software testing are *black box* and *white box* testing. White box testing involves checking that every path in a module has been tested and involves defining and executing test cases to ensure code and branch coverage. The goal of black box testing is to verify the functionality of a module or feature or the complete system itself. Testing is both a constructive activity in that it is verifying the correctness of functionality, and it may be a destructive activity in that the objective is to find defects in the implemented software. Testing verifies that the requirements are correctly implemented, and it yields the presence or absence of defects.

There are various types of testing including unit, system, performance, and usability testing. The effectiveness of the testing is influenced by the maturity of the test process employed. Testing is described in detail in the remainder of this book.

### 1.5.3 Software Quality Assurance

The software quality assurance department provides visibility into the quality of the work products being built and the processes being used to create them. Its activities include audits of the various groups involved in software development.

The quality group promotes quality in the organization and is independent of the development group. It provides an independent assessment of the quality of the product being built, and this viewpoint is independent of the project manager and development viewpoint. The quality assurance group acts as the voice of the customer and aims to ensure that quality is considered at each step in the process.

The quality group will perform audits of various projects, groups, and departments and will determine the extent to which the process is followed and report any weaknesses in the processes and non-compliances identified. Any non-compliance issues that are not addressed may be escalated to the next level of management for resolution. Its key responsibilities are:

- Promotes quality in organization
- Conducts audits to verify compliance
- Reports audit results to management

- Provides visibility to management on processes followed
- Facilitates software process improvement
- Release sign-offs.

The quality audit provides visibility into the work products and processes used to develop the work products. The audit consists of an interview with the project team, and the auditor examines the processes followed and deliverables produced by each team member and assesses if there are any quality risks associated with the project based on the information provided.

The *auditor* needs good written and verbal communication skills and will consider the role that the participant is performing and relates this to the defined process for their area. The auditor writes a report detailing the findings from the audit and the recommended corrective actions with respect to any identified non-compliance to the defined procedures. He/she will perform follow-up activity at a later stage to verify that the corrective actions have been carried out. The audit activities include planning activities, the audit meeting, gathering data, reporting the findings and assigning actions, and following the actions through to closure.

### 1.5.4 Problem-Solving Techniques

There is a relationship between the quality of the process and the quality of the products built from the process. Defects may be due to a defect in the process itself, and so it is important to identify the causes of defects and to correct any systemic defects in the process.

*Problem-solving teams* are formed to solve a particular problem and to identify appropriate corrective actions. The team may be disbanded after successful resolution of the problem, and they first agree on the problem to be solved. They collect and analyse the facts and perform analysis to determine the appropriate solution. They use various tools such as fishbone diagrams, histograms, trend charts, Pareto diagrams, and bar charts to assist with problem-solving and to analyse and identify appropriate corrective actions.

*Fishbone Diagrams*
This well-known cause-and-effect diagram is in the shape of the backbone of a fish. The approach is to identify the possible causes of some particular quality effect. These may include people, materials, methods, and timing. Each of the main causes may then be broken down into subcauses. The root cause is then identified, as often 80% of problems are due to 20% of causes (the 80:20 rule).

*Histograms*
A histogram is a way of representing data via a frequency distribution in bar chart format, and it is a graphical representation of the underlying distribution of the data. It illustrates the shape, variation, and centring of the underlying distribution. The data is divided into a number of buckets, where a bucket is a particular range of data

values, and the relative frequency of each bucket is displayed in bar format. The shape of the process and its spread from the mean is evident from the histogram.

*Pareto Chart*
The objective of a Pareto chart is to identify the key problems and to focus on these. Problems are classified into various types or categories, and the frequency of each category of problem is then determined. The chart is displayed in a descending sequence of frequency, with the most significant category detailed first, and the least significant category detailed last. The success in problem-solving activities over a period of time may be judged from the trends in the Pareto chart, and if problem-solving activities are successful, then the key problem categories in the old chart should show a noticeable improvement in the new Pareto chart.

*Trend Graph*
A trend graph is a graph of a variable over time and is a study of observed data for trends or patterns over time.

*Scatter Graphs*
The scatter diagram is used to measure the relationship between variables and to determine whether there is a correlation between the variables. The results may be a positive correlation, negative correlation, or no correlation between the data. The scatter diagram provides a means to confirm a hypothesis that two variables are related and provides a visual means to illustrate the potential relationship.

*Failure Mode Effect Analysis*
This involves identifying all of the possible failures of the system and the impact of each failure. Each possible failure mode is documented, as well as the impact of failure, the cause of failure, the frequency of occurrence, its severity, the estimate of detection of the failure, the risk and corrective action to minimize the risk. FMEAs are usually applied at the design stage.

Problem-solving techniques are discussed in more detail in Chap. 9.

### 1.5.5 Cost of Quality

Crosby argued that the most meaningful measurement of quality is the cost of quality and that the emphasis of the improvement activities should be to reduce the *cost of poor quality* (COPQ).

The cost of quality includes the cost of external and internal failure, the cost of providing an infrastructure to prevent the occurrence of problems, and the cost of the infrastructure to verify the correctness of the product. It was divided into four subcategories (Table 1.9) by Feigenbaum in the 1950s and evolved further by James Harrington of IBM.

**Table 1.9** Cost of quality categories

| Type of Cost | Description |
|---|---|
| Cost external failure | This includes the cost of external failure and includes engineering repair, warranties, and a customer support function |
| Cost internal failure | This includes the internal failure cost and includes the cost of reworking and retesting of any defects found internally |
| Cost prevention | This includes the cost of maintaining a quality system to prevent the occurrence of problems and includes the cost of software quality assurance and the cost of training |
| Cost appraisal | This includes the cost of verifying the conformance of a product to the requirements and includes the cost of provision of software inspections and testing processes |

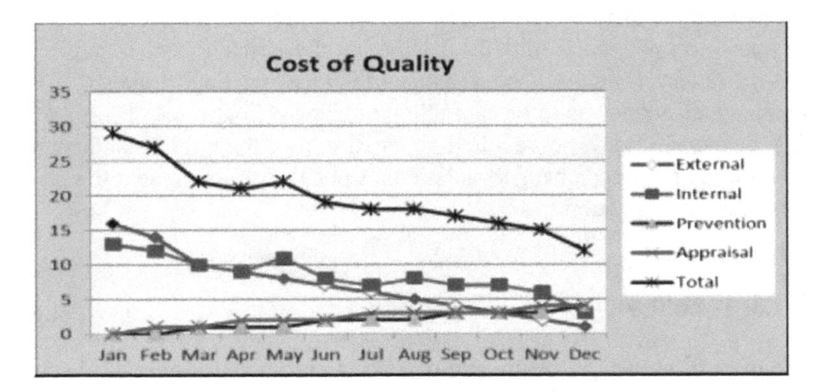


**Fig. 1.9** Cost of quality

The cost of quality graph (Fig. 1.9) will initially show high external and internal costs and very low prevention costs, and the total quality costs will be high. However, as an effective quality system is put in place and becomes fully operational there will be a noticeable decrease in the external and internal cost of quality and a gradual increase in the cost of prevention and appraisal. The total cost of quality will substantially decrease, as the cost of provision of the quality system is substantially below the savings gained from lower cost of internal and external failure.

### 1.5.6 Software Process Improvement

Software process improvement initiatives support the organization in achieving its key business goals such as delivering software faster to the market, improving quality, reducing or eliminating waste. The objective is to work smarter and to build software better, faster, and cheaper than competitors. It makes business sense and provides a tangible return on investment.

An improvement programme is a project in its own right and needs to be managed as such. Model-based approaches to process improvement involve using models such as the CMM, CMMI, ISO 9000, PSP or TSP. A software process maturity model provides a set of best practices in software engineering, and an assessment of the organization against the model will yield the current strengths and weaknesses of the organization with respect to the model. The organization needs to prioritize the improvements that will give the greatest business return.

The employees of the company are, in effect, the owners of the process infrastructure within the organization, as they work with the processes and procedures on a daily basis. They have an interest in having the best possible processes, and a good improvement programme will empower employees to make suggestions for continuous improvement. A reward and recognition mechanism helps to make process improvement part of the organization culture.

Improvement tends to be most successful when performed in small steps rather than trying to do too much initially. It is generally easier for an organization to adjust to a series of small changes rather than one big major change. Changes within an organization need to be carefully planned and controlled. Training for the existing employees may be required to ensure that they fully understand the rationale for the proposed changes and are in a position to implement the proposed changes in the organization.

### 1.5.7 Software Metrics

The use of measurement is an integral part of science and engineering disciplines, and software measures are increasingly used in software engineering. The term *"software metric"* was coined by Tom Gilb in his influential book on software measurement (Gilb 1977). The purpose of measurement in software engineering is to provide an objective indication of the effectiveness of the organization in achieving its key goals and objectives.

It is essential that the measurements are relevant and closely related to the organization goal. One way to ensure this is to employ the *goal, question, metric* (GQM) approach which mandates the organization to first identify its key goals; then, it identifies the questions which need to be answered to assess the extent to which the goal is being satisfied, and then, it formulates a metric to give an objective answer to the particular question. This approach was formulated by Victor Basili and others and is described in Basili and Rombach (1988).

Measurement may be used to verify that an organization has actually improved, as quantitative data before and after the improvement initiative can be compared to judge the extent of the improvements. The initial measurements prior to the improvement programme serve as the baseline measurement of the current capability of the organization. A successful improvement programs will lead to improvements, and this will be reflected in the metrics. The implementation of metrics involves:

- Business goals
- Questions related to goals
- Metrics
- Data gathering
- Presentation of charts
- Trends
- Action plans.

Metrics are discussed in more detail in Chap. 9.

### 1.5.8 Customer Satisfaction

The customer will ultimately judge the effectiveness of the quality management system in delivering high-quality software, and the level of customer satisfaction will influence the customer in purchasing again from the company or recommending the company. Customer satisfaction surveys are used to determine the level of customer satisfaction with the company.

A customer satisfaction survey involves the customer rating the organization in several key areas such as the quality of the software, its reliability, the timeliness of delivery, and so on. The process takes the form of a closed feedback loop, and the customer satisfaction feedback will be analysed and acted upon appropriately.

The survey is conducted, and the feedback analysed and used to prepare the action plan. The actions are executed, and the customer is surveyed again at later date (Fig. 1.10). The follow-up activity may involve a telephone conversation with the customer or a visit to the customer site to discuss the specific issues. The objective is to ensure that customers are totally satisfied with the product and service, as a loyal customer will repurchase and recommend the company to other potential customers.

**Fig. 1.10** Customer satisfaction process

**Table 1.10** Sample customer satisfaction questionnaire

| No | Question | Unacceptable | Poor | Fair | Satisfied | Excellent | N/A |
|---|---|---|---|---|---|---|---|
| 1. | Quality of software | □ | □ | □ | □ | □ | □ |
| 2. | Ability to meet agreed dates | □ | □ | □ | □ | □ | □ |
| 3. | Timeliness of projects | □ | □ | □ | □ | □ | □ |
| 4. | Effective testing of software | □ | □ | □ | □ | □ | □ |
| 5. | Expertise of staff | □ | □ | □ | □ | □ | □ |
| 6. | Value for money | □ | □ | □ | □ | □ | □ |
| 7. | Quality of support | □ | □ | □ | □ | □ | □ |
| 8. | Ease of installation | □ | □ | □ | □ | □ | □ |
| 9. | Ease of use | □ | □ | □ | □ | □ | □ |
| 10. | Timely problem resolution | □ | □ | □ | □ | □ | □ |

The customer satisfaction process is summarized as follows:

- Define customer surveys
- Send customer surveys
- Customer satisfaction ratings
- Customer meeting and key issues
- Action plans and follow-up
- Metrics for customer satisfaction.

The questionnaire will vary according to the business, but it will cover the relevant questions to determine where the organization is weak and areas where it is strong. The questions typically employ a rating scheme to allow the customer to give quantitative feedback on satisfaction, and the survey will also enable the customer to go into more detail on issues.

A sample survey form with 10 questions is included in Table 1.10, and the form will also include open-ended questions to enable the customer to go into more detail on any issues. Customer satisfaction metrics provide visibility into the level of customer satisfaction and enable trends to be determined (Fig. 1.11).

### 1.5.9 Assessments (Appraisals)

The objective of an assessment (or *appraisal*) of an organization is to determine its maturity with respect to a maturity model such as the CMMI or SPICE or against an international quality standard such as ISO 9000.

The appraisal is performed by an external or internal assessment team and yields the strengths and weaknesses of the organization with respect to the model. The appraisal report is used to plan and prioritize future improvements.

It is a major review of the organization, and it needs to be conducted by an experienced assessment team. It involves interviews with the project managers and

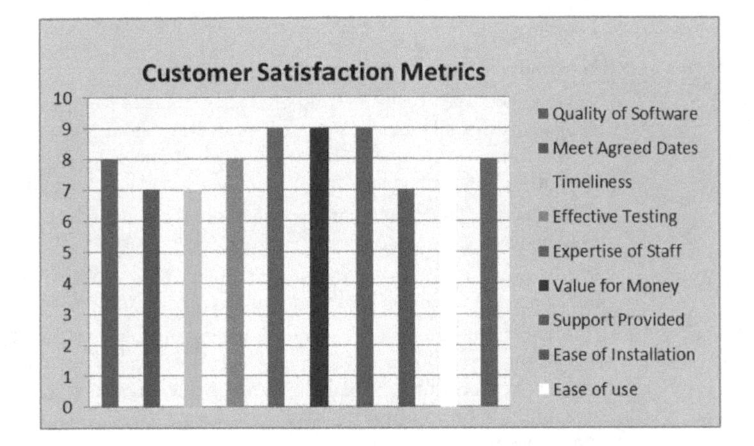


**Fig. 1.11** Customer satisfaction metrics

project teams as well as the review of relevant documentation. The assessment report will detail the extent to which the model is implemented, and any gaps and improvement opportunities are highlighted in the report.

### 1.5.10 Total Quality Management

Total quality management (TQM) is a management philosophy that is focused on quality and on developing a culture of quality in the organization. It is a holistic approach, and it applies to all levels and functions within the organization. Quality is a company-wide objective, and the goal is total customer satisfaction. The company aims to deliver products and services that totally satisfy the customer needs.

TQM uses many of the ideas of the pioneers in the quality movement. Management is required to take charge of the implementation of quality management, and all staff will need to be trained in quality improvement activities.

The implementation of TQM involves a focus on all areas in the organization and in identifying potential improvements. The problems in the particular area are evaluated, and data is collected and analysed. An action plan is then prepared and the actions implemented and monitored. This is then repeated for continuous improvement. The implementation is summarized as follows:

- Identify improvement area
- Problem evaluation
- Data collection
- Data analysis
- Action plan
- Implementation of actions

**Table 1.11** Total quality management

| Part | Description |
|---|---|
| Customer focus | This involves identifying internal and external customers and recognizing that all customers have expectations and rights which need to be satisfied first time and every time. Quality must be considered in every aspect of the business, and the focus is on fire prevention |
| Process | This involves a focus on the process and improvement to the process via problem-solving to reduce waste and eliminate errors |
| Measurement and analysis | This involves setting up a measurement programme to enable objective and effective analysis of the quality of the process and product |
| Human factors | This involves developing a culture of quality and customer satisfaction throughout the organization. The core values of quality and customer satisfaction need to be instilled in the organization. This requires training for the employees on quality, customer satisfaction, and continuous improvement |

- Monitor effectiveness
- Repeat.

There are four main parts of TQM which are summarized in Table 1.11.

The ISO 9000 standard [see Chap. 11 of O'Regan (2014)] is a structured approach to the implementation of TQM. Its clauses are guidelines for what needs to be done and include requirements to be satisfied for the organization to satisfy ISO 9000.

## 1.6 Miscellaneous

Software quality management is, in many ways, the application of common sense to software engineering. In this section, we discuss organization culture and change as well as legal aspects of failure, and finally, we discuss quality and the Web.

### 1.6.1 Organization Culture and Change

Every organization has a distinct culture, and this reflects the way in which things are done in the company. Organization culture includes the ethos of the organization, its core values, its history, its success stories, its people, amusing incidents, and so on. The culture of the organization may be favourable or unfavourable to developing high-quality software.

Occasionally, a change to the organization culture is required, and this may be difficult as it could involve changing its fundamental ways of working, and there may be a resistance to this. Successful change management often involves:

- Kick-off meeting
- Motivate rationale for changes
- Present plan
- Training
- Implement changes
- Monitor implementation
- Institutionalize.

The culture of an organization is often illustrated by the phrase: *"That's the way we do things around here"*. For example, the evolution from one level of the CMM to another often involves a change the way that things are done in the organization. The focus on prevention requires a change in mindset to focus on *problem-solving* and *fire prevention*, rather than on *firefighting*.

### 1.6.2 Law of Negligence

The impact of a flaw in software may be catastrophic, and several software failures were discussed earlier in this chapter. Clearly, every organization must take all reasonable precautions to prevent the occurrence of defects, especially in the safety-critical domain where defects may cause major damage or even loss of life. Reasonable precautions consist of having appropriate software engineering practices in place to allow the organization to consistently produce high-quality software.

A quality management system indicates that the organization takes software quality seriously and that has a sound software development process in place that serves the needs of the organization and its customers. Modem quality assurance systems include processes for software inspections, testing, quality audits, customer satisfaction, software development, project planning, etc.

The organization will require evidence or records to prove that the quality management system is in place that it is appropriate for the organization and that it is fully operational within the organization. The proof that the quality system is actually operational typically takes the form of records of the various activities. The records also enable the organization to prepare a legal defence to show that it took all reasonable precautions in software development, especially if a customer decides to take legal action for negligence against the software provider following a serious problem in the software at the customer site.

The presence of records may be used to indicate that all reasonable steps were taken, and the records typically include lists of all the deliverables in the project; minutes of project meetings; records of reviews of requirements, design, and software code, records of test plans and test results; and so on.

### 1.6.3 Quality and the Web

The explosive growth of the World Wide Web and electronic commerce has made the quality of websites a key concern. Web technology is rapidly becoming ubiquitous in society and is quite distinct from other software systems in that:

- It may be accessed from anywhere in the world
- It may be accessed by many different browsers
- The usability and look and feel of the application is a key concern
- The performance of the website is a key concern
- Security is a key concern
- The website must be capable of dealing with a large number of transactions at any time
- The website has very strict availability constraints (typically 24 $\times$ 365)
- The website needs to be highly reliable.

It is inappropriate to employ the waterfall lifecycle for this domain, and usually a spiral lifecycle will be employed as the requirements are often incomplete at project initiation and evolve to the agreed set during the project. Often, rapid application development (RAD), joint application development (JAD) or the Agile methodology is employed.

## 1.7 Review Questions

1. Discuss the contributions of Deming and Juran.
2. Describe Crosby's maturity grid and discuss how it influenced the Capability Maturity Model?
3. Explain why Watts Humphrey is considered the father of software quality.
4. Explain the difference between software inspections and testing?
5. What is an assessment (appraisal) and explain how it forms part of the improvement cycle.
6. Why is the cost of poor quality an important measure?
7. Discuss the role of software metrics in problem-solving.
8. Explain the importance of customer satisfaction and describe how it may be measured.

## 1.8 Summary

This chapter gave a short introduction to the software quality field, and we discussed the contributions of several pioneers in the quality field including Shewhart, Deming, Juran, and Crosby. We also discussed Watts Humphrey, who is considered the father of software quality.

We examined various definitions of quality such as Crosby's "conformance to the requirements" and Juran's "fitness for purpose", as well as considering the various dimensions of software product quality listed in ISO 9126.

We considered several software failures such as the Ariane 5 disaster, the year 2000 problem, and a maths bug in the Intel Pentium microprocessor. A software failure may have devastating consequences and so it is essential to develop high-quality software.

We discussed software inspections that build quality into the software; software testing that verifies that the software is of high quality as well as finding defects in the software; software quality assurance to provide visibility into the processes; problem-solving techniques to prevent problems from re-occurring; the cost of poor quality to the organization; software process improvement to improve the key processes in the organization; and customer satisfaction to determine the level of customer satisfaction with the organization.

## References


Basili V, Rombach H (1988) The TAME project. Towards improvement-oriented software environments. IEEE Trans Softw Eng 14(6)

Brooks F (1975) The mythical man month. Addison Wesley, Boston

Brooks F (1986) No silver bullet. Essence and accidents of Software Engineering. Information processing. Elsevier, Amsterdam

Crosby P (1979) Quality is free. The art of making quality certain. McGraw Hill, New York

Deming WE (1986) Out of crisis. M.I.T. Press, Cambridge

Fagan M (1976) Design and code inspections to reduce errors in software development. IBM Syst J 15(3)

Gilb T (1977) Software metrics. Winthrop Publishers, Winthrop

Humphry W (1989) Managing the software process. Addison Wesley, Boston

ISO/IEC 9126 (1991) Information Technology. Software product evaluation: quality characteristics and guidelines for their use

Juran J (1951) Juran's quality handbook. McGraw Hill, New York

Lions JL (1996) Ariane 5. Flight 501. Failure report by enquiry board

Manley E (1995) Taurus: how I lived to tell the tale (American Programmer: Software failures)

O'Regan G (2014) Introduction to software quality. Springer, Berlin

Shewhart W (1931) The economic control of manufactured products. D. van Nostrand & Co. Inc., New York

Standish Group Research Note (1999) Estimating: art or science. Featuring Morotz cost expert

# Fundamentals of Software Engineering

# 2

**Key Topics**

Standish report
Software lifecycles
Waterfall model
Spiral model
Rational unified process
Agile development
Software inspections
Software testing
Project management

## 2.1 Introduction

The approach to software development in the 1950s and 1960s has been described as the "*Mongolian Hordes Approach*" by Brooks (1975).[1] The "method" was applied to projects that were running late, and it involved adding a large number of programmers to the project, with the expectation that this would enable the project schedule to be recovered. However, this approach was deeply flawed as it led to inexperienced programmers with inadequate knowledge of the project attempting to

[1]Brooks was the project manager for the IBM System 360 project. The "Mongolian Hordes" management myth is the belief that adding more programmers to a software project that is running late will allow catchup. Brooks confirmed that adding people to a late software project actually makes it later.



G. O'Regan, *Concise Guide to Software Testing*,
Undergraduate Topics in Computer Science,
https://doi.org/10.1007/978-3-030-28494-7_2

solve problems, and they inevitably required significant time from the other project team members.

This resulted in the project being delivered even later, as well as subsequent problems with quality (i.e. the approach of throwing people at a problem does not work). The approach to software development back in the 1950/60s was characterized by the philosophy:

*The completed code will always be full of defects.*
*The coding should be finished quickly to correct these defects.*
*Design as you code approach.*

This philosophy accepted defeat in software development, and suggested that irrespective of a solid engineering approach, that the completed software would always contain lots of defects, and that it therefore made sense to code as quickly as possible, and to then identify the defects so as to correct them as quickly as possible.

In the late 1960s, it was clear that the existing approaches to software development were deeply flawed, and that there was an urgent need for change. The NATO Science Committee organized two famous conferences to discuss critical issues in software development (Naur and Randell 1975). The first conference was held at Garmisch, Germany, in 1968, and it was followed by the second conference in Rome in 1969. Over fifty people from eleven countries attended the Garmisch conference, including Peter Naur who produced a report on the conference (Naur and Randell 1969). The NATO conferences highlighted problems that existed in the software sector in the late 1960s, and the term "*software crisis*" was coined to refer to these. There were problems with budget and schedule overruns, as well as the quality and reliability of the delivered software.

The conference led to the birth of *software engineering* as a discipline in its own right, and the realization that programming is quite distinct from science to mathematics. Programmers are like engineers in that they build software products, and they therefore need education in traditional engineering as well as on the latest technologies. The education of a classical engineer includes product design and mathematics. However, often computer science education places an emphasis on the latest technologies, rather than on the important engineering foundations of designing and building high-quality products that are safe for the public to use.

Programmers, therefore, need to learn the key engineering skills to enable them to build products that are safe for the public to use. This includes a solid foundation on design and on the mathematics required for building safe software products. Mathematics plays a key role in classical engineering, and in some situations it may also assist software engineers in the delivery of high-quality software products. Several mathematical approaches to assist software engineers are described in O'Regan (2006, 2017b).

There are parallels between the software crisis in the late 1960s and serious problems with bridge construction in the nineteenth century. Several bridges collapsed, or were delivered late or overbudget, due to the fact that people involved in

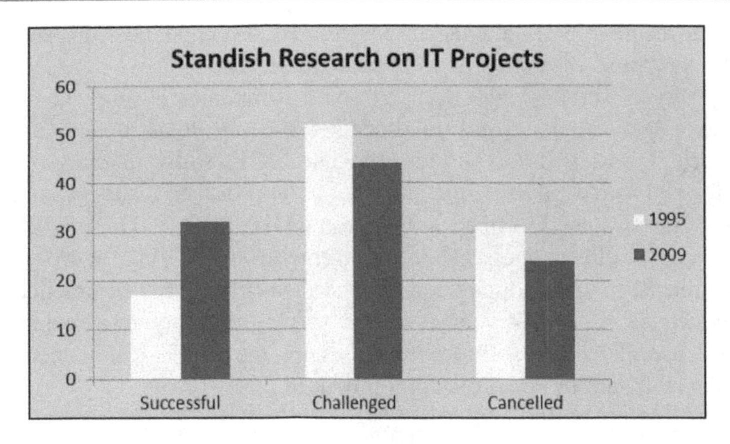


**Fig. 2.1** Standish report—results of 1995 and 2009 Survey

their design and construction did not have the required engineering knowledge. This led to poorly designed and constructed bridges that later collapsed leading to loss of life, as well as endangering the lives of the public.

This led to legislation requiring engineers to be licenced by the Professional Engineering Association prior to practicing as engineers. This organization specified a core body of knowledge that the engineer is required to possess, and the licensing body verifies that the engineer has the required qualifications and experience. This helps to ensure that only personnel competent to design and build products actually do so. Engineers have a professional responsibility to ensure that the products are properly built and are safe for the public to use.

The Standish group has conducted research (Fig. 2.1) on the extent of problems with IT projects since the mid-1990s. These studies were conducted in the USA, but there is no reason to believe that European or Asian companies perform any better. The results indicate serious problems with on-time delivery of projects and projects being cancelled prior to completion.[2] However, the comparison between 1995 and 2009 suggests that there have been improvements with a greater percentage of projects being delivered successfully and a reduction in the percentage of projects being cancelled.

Fred Brooks argues that software is inherently complex, and that there is no *silver bullet* that will resolve all of the problems associated with software development such as schedule or budget overruns (Brooks 1975, 1986). Poor software quality may adversely impact the customer and even cause loss of life. It is, therefore, essential that software development organizations place sufficient emphasis on quality throughout the software development lifecycle.

The Y2K problem was caused by a two-digit representation of dates, and it required major rework of legacy software for the new millennium. Clearly, well-designed programs would have hidden the representation of the date, which

[2]These are IT projects covering diverse sectors including banking, telecommunications, etc., rather than pure software companies. Software companies following maturity frameworks such as the CMMI generally achieve more consistent results.

would have required minimal changes for year 2000 compliance. Instead, companies spent vast sums of money to rectify the problem.

The quality of software produced by some companies is quite good.[3] These companies employ mature software processes and are committed to continuous improvement. There is a lot of industrial interest in software process maturity models for software organizations, and various approaches to assess and mature software companies are described in O'Regan (2010, 2014).[4] These models focus on improving the effectiveness of the management, engineering, and organization processes related to software engineering and in introducing best practice in software engineering. It is a key tenet of the software quality movement that the disciplined use of mature software processes by the software engineers enables high-quality software to be consistently produced.

## 2.2 What Is Software Engineering?

Software engineering involves the multi-person construction of multi-version programs. The IEEE 610.12 definition is as follows:

> Software engineering is the application of a systematic, disciplined, quantifiable approach to the development, operation, and maintenance of software; that is, the application of engineering to software, and the study of such approaches.

Software engineering includes as follows:

1. Methodologies to design, develop, and test software to meet customers' needs.
2. Software is engineered. That is, the software products are properly designed, developed, and tested in accordance with sound engineering principles.
3. Quality and safety are properly addressed.
4. Mathematics may be employed to assist with the design and verification of software products. The level of mathematics employed will depend on the *safety-critical* nature of the product. Systematic peer reviews and rigorous testing will often be sufficient to build quality into the software, with heavy *mathematical techniques reserved for safety and security critical software.*
5. Sound project management and quality management practices are employed.
6. Support and maintenance of the software are properly addressed.

[3]I recall projects at Motorola that regularly achieved 5.6σ-quality in a L4 CMM environment (i.e. approx. 20 defects per million lines of code. This represents very high quality).

[4]Approaches such as the CMM or SPICE (ISO 15504) focus mainly on maturing management and organizational practices used in software development. The emphasis is on defining software processes that are fit for purpose and consistently following them. The process maturity models focus on what needs to be done rather how it should be done. This gives the organization the freedom to choose the appropriate implementation to meet its needs. The models provide useful information on practices to consider in the implementation.

Software engineering is not just programming. It requires the engineer to state precisely the requirements that the software product is to satisfy and then to produce designs that will meet these requirements. The project needs to be planned and delivered on time and budget. The requirements must provide a precise description of the problem to be solved, i.e. *it should be evident from the requirements what is and what is not required.*

The requirements need to be rigorously reviewed to ensure that they are stated clearly and unambiguously and reflect the customer's needs. The next step is then to create the design that will solve the problem, and it is essential to validate the correctness of the design. Next, the software code to implement the design is written, and peer reviews and software testing are employed to verify and validate the correctness of the software.

The verification and validation of the design are rigorously performed for safety-critical systems, and it is sometimes appropriate to employ mathematical techniques for these systems. However, it will often be sufficient to employ peer reviews (software inspections) and testing as these methodologies provide a high degree of rigour. These may include methodologies such as Fagan inspections (Fagan 1976), Gilb inspections (Gilb and Graham 1994), or Prince 2's approach to quality reviews (Office of Government Commerce 2004).

The term "*engineer*" is a title that is awarded on merit in classical engineering. It is generally applied only to people who have attained the necessary education and competence to be called engineers and who base their practice on sound engineering principles. The title places responsibilities on its holder to behave professionally and ethically. Often in computer science, the term "*software engineer*" is employed loosely to refer to anyone who builds things, rather than to an individual with a core set of knowledge, experience, and competence.

Several computer scientists (e.g. Parnas[5]) have argued that computer scientists should be educated as engineers to enable them to apply appropriate scientific principles to their work. They argue that computer scientists should receive a solid foundation in mathematics and design, to enable them to have the professional competence to perform as engineers in building high-quality products that are safe for the public to use. The use of mathematics is an integral part of the engineer's work in other engineering disciplines, and so the *software engineer* should be able to use mathematics to assist in the modelling or understanding of the behaviour or properties of the proposed software system.

Software engineers need education[6] on specification, design, and turning designs into programs, software inspections, and testing. The education should enable the software engineer to produce well-structured programs that are fit for purpose. Parnas has argued that software engineers have responsibilities as professional

[5]Parnas advocates a solid engineering approach with the extensive use of classical mathematical techniques in software development. He also introduced information hiding in the 1970s, which is now a part of object-oriented design.

[6]Software companies that are following approaches such as the CMM or ISO 9001 considering the education and qualification of staff prior to assigning staff to performing specific tasks.

engineers.[7] They are responsible for designing and implementing high-quality and reliable software that is safe to use. They are also accountable for their decisions and actions[8] and have a responsibility to object to decisions that violate professional standards.

Engineers are required to behave professionally and ethically with their clients. The membership of the professional engineering body requires the member to adhere to the code of ethics[9] of the profession. Engineers in other professions are licenced, and therefore Parnas argues that a similar licencing approach be adopted for professional software engineers[10] to provide confidence that they are competent for the particular assignment. Professional software engineers are required to follow best practice in software engineering and the defined software processes.[11]

Many companies invest heavily in training, as educated and knowledgeable employees are essential in the delivery of high-quality products and services. Employees need to receive appropriate training related to the roles that they perform, such as project management, software design and development, software testing, and service management. The fact that the employees are professionally qualified increases the confidence in the ability of the company to deliver high-quality products and services. A company that pays little attention to the competence and continuous development of its staff will achieve poor results and suffer a loss of reputation and market share.

[7]The ancient Babylonians were familiar with the concept of accountability, and they employed a code of laws (known as the Hammurabi Code) c. 1750 B.C. It included a law that stated that if a house collapsed and killed the owner, then the builder of the house would be executed.

[8]It is unlikely that an individual programmer would be subject to litigation in the case of a flaw in a program causing damage or loss of life. A comprehensive disclaimer of responsibility for problems rather than a guarantee of quality accompanies most software products. Software engineering is a team-based activity involving many engineers in various parts of the project, and it would be potentially difficult for an outside party to prove that the cause of a particular problem is due to the professional negligence of a particular software engineer, as there are many others involved in the process such as reviewers of documentation and code and the various test groups. Companies are more likely to be subject to litigation, as a company is legally responsible for the actions of their employees in the workplace, and a company is a wealthier entity than one of its employees. The legal aspects of licencing software may protect software companies from litigation. However, greater legal protection for the customer can be built into the contract between the supplier and the customer for bespoke software development.

[9]Today, many software companies have a defined code of ethics that employees are expected to adhere. Larger companies will wish to project a good corporate image to be respected worldwide.

[10]The British Computer Society (BCS) has introduced a qualification system for computer science professionals, which is used to show that professionals are properly qualified. The most important of these is the BCS Information Systems Examination Board (ISEB) which allows IT professionals to be qualified in service management, project management, software testing, and so on.

[11]Software companies that are following the CMMI or ISO 9001 standards will employ audits to verify that the processes and procedures have been followed. Auditors report their findings to management, and the findings are addressed appropriately by the project team and affected individuals.

## 2.3 Challenges in Software Engineering

The challenge in software engineering is to deliver high-quality software on time and on budget to customers. We discussed the research done by the Standish Group earlier in this chapter, and their 1998 research (Fig. 1.1) on project cost overruns in the USA indicated that 33% of projects are between 21 and 50% over estimate, 18% are between 51 and 100% over estimate, and 11% of projects are between 101 and 200% overestimate.

The accurate estimation of project cost, effort, and schedule is a challenge in project management. Therefore, project managers need to determine how good their estimation process actually is and to make appropriate improvements. Software metrics provide an objective way to see improvements in estimation from a reduced variance between estimated and actual effort. The project manager will report the actual versus estimated effort and schedule for the project.

Risk management is an important part of project management, and the objective is to identify potential risks early and throughout the project and to manage them appropriately. The probability of each risk occurring and its impact is determined, and the risks are managed during project execution.

Software quality needs to be properly planned to enable the project to deliver a quality product. Flaws with poor quality software lead to a negative perception of the company and may potentially lead damage to the customer relationship with a subsequent loss of market share.

There is a strong economic case to building quality into the software, as less time is spent in reworking defective software. The cost of poor quality (COPQ) should be measured and targets set for its reductions. It is important that lessons are learned during the project and acted upon appropriately. This helps to promote a culture of continuous improvement.

Several high-profile software failures were discussed in Chap. 1. These include the millennium bug (Y2K) problem, the floating-point maths bug in the Intel Pentium microprocessor in the early 1990s, the European Space Agency Ariane-5 disaster, and so on. These failures led to the embarrassment for the organizations involved, as well as the associated cost of replacement and correction.

The millennium bug was due to the use of two digits to represent dates rather than four digits. The solution involved finding and analysing all code that had a Y2K impact, planning and making all necessary changes, and verifying the correctness of the changes made. The worldwide cost of correcting the millennium bug is estimated to have been in billions of dollars.

The Intel Corporation was slow to acknowledge the floating-point problem in its Pentium microprocessor and in providing adequate information on its impact to its customers. It incurred a large financial cost in replacing microprocessors for its customers as well as damage to its reputation. The Ariane 5 failure caused major embarrassment and damage to the credibility of the European Space Agency (ESA). Its maiden flight ended in failure on June 4, 1996, after a flight time of just 40 s.

These failures indicate that the quality needs to be carefully considered when designing and developing software. The effect of software failure may be large costs to correct the software, loss of credibility of the company, or even loss of life.

## 2.4 Software Processes and Lifecycles

Organizations vary by size and complexity, and the processes they employ will reflect the nature of their business. The development of software involves many processes such as those for defining requirements, processes for project estimation and planning, processes for design, implementation, testing, and so on.

It is important that the processes are fit for purpose, and a key premise in the software quality field is that the quality of the resulting software is closely influenced by the quality and maturity of the underlying processes and compliance to them. Therefore, it is necessary to focus on the quality of the processes as well as the quality of the resulting software.

There is, of course, little point in having high-quality processes unless the employees in the organization use them consistently. This requires that the employees are trained on the processes, and that process discipline is instilled into an appropriate audit strategy. The software process assets in an organization generally consist of as follows:

- A software development policy for the organization
- Process maps that describe the flow of activities
- Procedures and guidelines that describe the processes in more detail
- Checklists to assist with the performance of the process
- Templates for the performance of specific activities (e.g. design, testing)
- Training materials.

The processes used to develop high-quality software generally include as follows:

- Project management process
- Requirements process
- Design process
- Coding process
- Peer review process
- Testing process
- Supplier selection and management processes
- Configuration management process
- Audit process
- Measurement process
- Improvement process
- Customer support and maintenance processes

There are several well-known lifecycles employed such as the waterfall model (Royce 1970), the spiral model (Boehm 1988), the Rational Unified Process (Rumbaugh 1999), and the Agile methodology (Beck 2000). The choice of a particular software development lifecycle for is determined from the needs of the specific project. The various lifecycles are described in more detail in the following sections.

### 2.4.1 Waterfall Lifecycle

The waterfall model (Fig. 2.2) starts with requirements gathering and definition. It is followed by the system specification (of the functional and non-functional requirements), the design and implementation of the software, and comprehensive testing. The testing generally includes unit, system, and user acceptance testing.

The waterfall model is employed for projects where the requirements can be identified early in the project lifecycle or are known in advance. We are treating the waterfall model as the "V" lifecycle model, with the left-hand side of the "V" detailing requirements, specification, design, and coding and the right-hand side detailing unit tests, integration tests, system tests and acceptance testing. Each phase has entry and exit criteria that must be satisfied before the next phase commences. There are several variations to the waterfall model.

Many companies employ a set of templates to enable the activities in various phases to be consistently performed. Templates may be employed for project planning and reporting, requirements definition, design, testing, and so on. These templates may be based on the IEEE standards or industrial best practice.

### 2.4.2 Spiral Lifecycles

The spiral model (Fig. 2.3) was developed by Barry Boehm in the 1980s (Boehm 1988), and it is useful for projects where the requirements are not fully known at

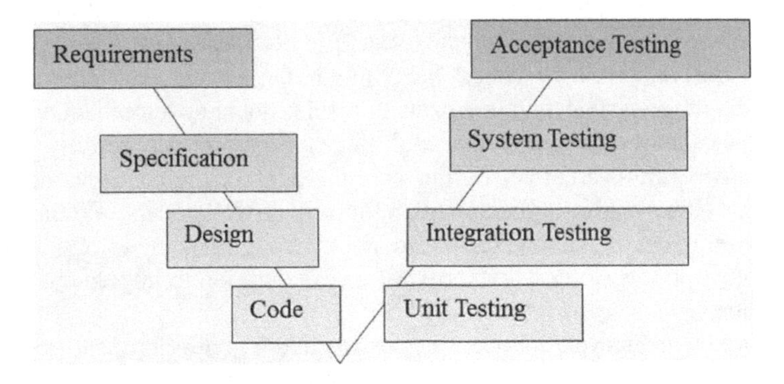


**Fig. 2.2** Waterfall V lifecycle model

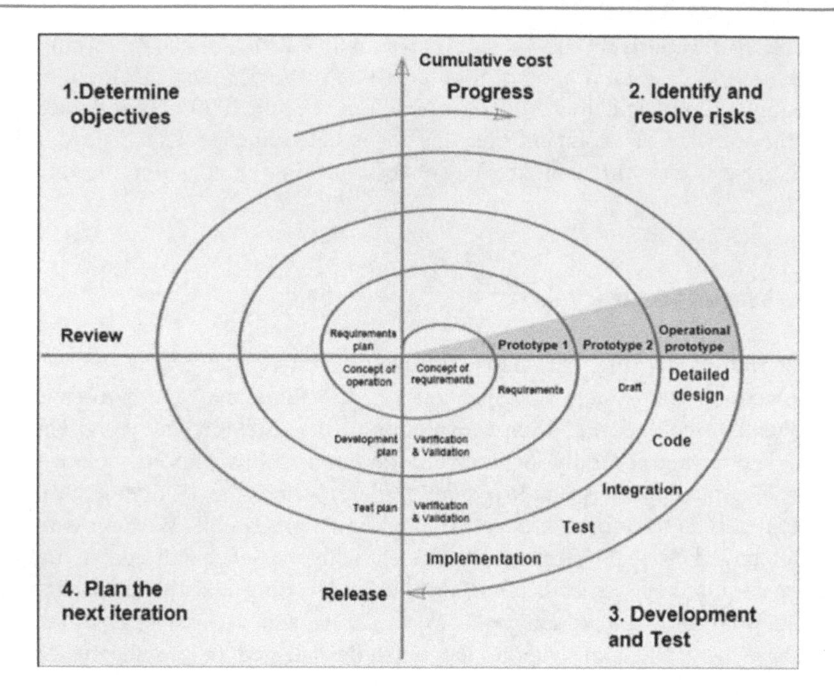


**Fig. 2.3** SPIRAL lifecycle model. Public domain

project initiation, or where the requirements evolve as a part of the development lifecycle. The development proceeds in a number of spirals, where each spiral implements new functionality and typically involves determining objectives and analysing the risks, updates to the requirements, design, code, testing, and a user review of the particular iteration or spiral.

The spiral is, in effect, a reusable prototype with the business analysts and the customer reviewing the current iteration, and providing feedback to the development team. The feedback is analysed and used to plan the next iteration. This approach is often used in joint application development, where the usability and look and feel of the application are a key concern. This is important in web-based development and in the development of a graphical user interface (GUI). The implementation of part of the system helps in gaining a better understanding of the requirements of the system, and this feeds into subsequent development cycles. The process repeats until the requirements and the software product are fully complete.

There are several variations of the spiral model including rapid application development (RAD), joint application development (JAD) models, and the dynamic systems development method (DSDM) model. The Agile methodology is popular, and it employs sprints (or iterations) of 2–4 weeks duration to implement a number of user stories.

There are other lifecycle models such as the iterative development process that combines the waterfall and spiral lifecycle model. The Cleanroom methodology was developed by Harlan Mills at IBM and includes a phase for formal specification. Its approach to software testing is based on the predicted usage of the software product, which allows a software reliability measure to be calculated. The Rational Unified Process (RUP) was developed by IBM Rational.

### 2.4.3 Rational Unified Process

The *Rational Unified Process* (Rumbaugh 1999) was developed at the Rational Corporation (now part of IBM) in the late 1990s. It uses the Unified Modelling Language (UML) as a tool for specification and design. UML is a visual modelling language for software systems that provides a means of specifying, constructing, and documenting the object-oriented system. James Rumbaugh, Grady Booch, and Ivar Jacobson developed it to facilitate the understanding of the architecture and complexity of a system.

RUP is *use case driven*, *architecture centric*, *iterative* and *incremental*, and it includes cycles, phases, workflows, risk mitigation, quality control, project management, and configuration control (Fig. 2.4). Software projects may be very complex, and there are risks that requirements may be incomplete, or that the interpretation of a requirement may differ between the customer and the project team. RUP is a way to reduce risk in software engineering.

Requirements are gathered as use cases, where the *use cases describe the functional requirements from the point of view of the user of the system.* They describe what the system will do at a high level and ensure that there is an appropriate focus on the user when defining the scope of the project. *Use cases also drive the development process*, as the developers create a series of design and implementation models that realize the use cases. The developers review each successive model for conformance to the use-case model, and the test team verifies that the implementation correctly and implements the use cases.

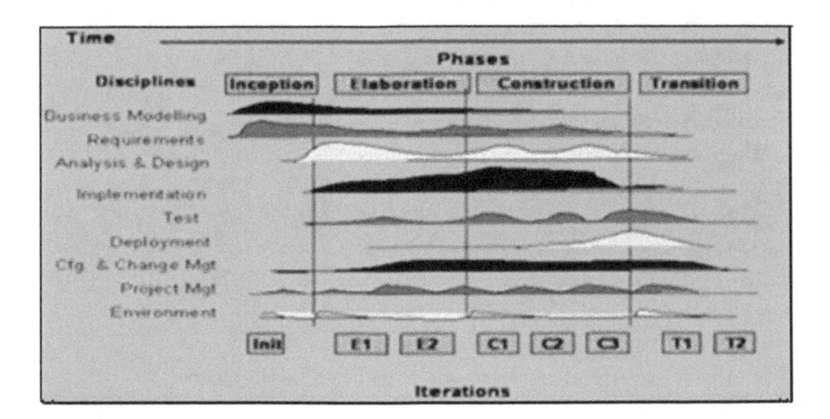


**Fig. 2.4** Rational unified process

The software architecture concept embodies the most significant static and dynamic aspects of the system. The architecture grows out of the use cases and factors such as the platform that the software is to run on, deployment considerations, legacy systems, and the non-functional requirements.

RUP decomposes the work of a large project into smaller slices or mini-projects, and *each mini-project is an iteration that results in an increment to the product.* The iteration consists of one or more steps in the workflow and generally leads to the growth of the product. If there is a need to repeat an iteration, then all that is lost is the misdirected effort of one iteration, rather that the entire product. Another words—RUP is a way to mitigate risk in software engineering.

### 2.4.4 Agile Development

There has been a massive growth of popularity among software developers in lightweight methodologies such as *Agile*. This is a software development methodology that is more responsive to customer needs than traditional methods such as the waterfall model. *The waterfall development model is similar to a wide and slow moving value stream*, and halfway through the project 100% of the requirements are typically 50% done. *However, for Agile development, 50% of requirements are typically 100% done halfway through the project.* Agile has a strong collaborative style of working including:

- Aims to achieve a narrow fast flowing value stream
- Feedback and adaptation employed in decision making
- User stories and sprints are employed
- Stories are either done or are not done (no such thing as 50% done)
- Iterative and incremental development is employed
- A project is divided into iterations
- An iteration has a fixed length (i.e. Time boxing is employed)
- Entire software development lifecycle is employed for the implementation of each story
- Change is accepted as a normal part of life in the Agile world
- Delivery is made as early as possible.
- Maintenance is seen as part of the development process
- Refactoring and evolutionary design employed
- Continuous integration is employed
- Short cycle times
- Emphasis on quality
- Stand-up meetings
- Plan regularly
- Direct interaction preferred over documentation
- Rapid conversion of requirements into working functionality
- Demonstrate value early
- Early decision making.

*Ongoing changes to requirements are considered normal in the Agile world*, and it is believed to be more realistic to change requirements regularly throughout the project rather than attempting to define all of the requirements at the start of the project. The methodology includes controls to manage changes to the requirements, and good communication and early regular feedback are an essential part of the process.

*A story may be a new feature or a modification to an existing feature*. It is reduced to the minimum scope that can deliver business value, and a feature may give rise to several stories. Stories often build upon other stories, and the entire software development lifecycle is employed for the implementation of each story. *Stories are either done or not done*, i.e. *there is such thing as a story being* 80% *done*. The story is complete only when it passes its acceptance tests. Stories are prioritized based on a number of factors including:

- Business value of story
- Mitigation of risk
- Dependencies on other stories.

The Scrum approach is an Agile method for managing iterative development, and it consists of an outline planning phase for the project followed by a set of sprint cycles (where each cycle develops an increment). *Sprint planning* is performed before the start of the iteration, and stories are assigned to the iteration to fill the available time. Each scrum sprint is of a fixed length (usually 2–4 weeks), and it develops an increment of the system. The estimates for each story and their priority are determined, and the prioritized stories are assigned to the iteration. *A short morning stand-up meeting is held daily* during the iteration, and attended by the scrum master, the project manager[12] and the project team. It discusses the progress made the previous day, problem reporting and tracking, and the work planned for the day ahead. A separate meeting is held for issues that require more detailed discussion.

Once the iteration is complete, the latest product increment is demonstrated to an audience including the product owner. This is to receive feedback and to identify new requirements. The team also conducts a retrospective meeting to identify what went well and what went poorly during the iteration. This is for continuous improvement of the future iterations. Planning for the next sprint then commences. The scrum master is a facilitator who arranges the daily meetings and ensures that the scrum process is followed. The role involves removing roadblocks so that the team can achieve their goals and communicating with other stakeholders.

*Agile employs pair programming and a collaborative style of working with the philosophy that two heads are better than one*. This allows multiple perspectives in decision making and a broader understanding of the issues.

[12]Agile teams are self-organizing and the project manager role is generally not employed for small projects (<20 staff).

Software testing is very important, and Agile generally employs automated testing for unit, acceptance, performance, and integration testing. Tests are run frequently with the goal of catching programming errors early. They are generally run on a separate build server to ensure that all dependencies are checked. Tests are rerun before making a release. *Agile employs test-driven development with tests written before the code*. The developers write code to make a test pass with ideally developers only coding against failing tests. This approach forces the developer to write testable code.

*Refactoring is employed in Agile as a design and coding practice*. The objective is to change how the software is written without changing what it does. Refactoring is a tool for evolutionary design where the design is regularly evaluated, and improvements are implemented as they are identified. It helps in improving the maintainability and readability of the code and in reducing complexity. The automated test suite is essential in showing that the integrity of the software is maintained following refactoring.

Continuous integration allows the system to be built with every change. Early and regular integration allows early feedback to be provided. It also allows all of the automated tests to be run thereby identifying problems earlier. Agile is discussed in more detail in Chap. 18 of O'Regan (2017a).

## 2.5 Activities in Waterfall Lifecycle

The waterfall software development lifecycle consists of various activities including:

- User (business) requirements definition
- Specification of system requirements
- Design
- Implementation
- Unit testing
- System testing
- UAT testing
- Support and maintenance

These activities are discussed in the following sections, and the description is specific to the non-Agile world.

### 2.5.1 User Requirements Definition

The user (business) requirements specify what the customer wants and define what the software system is required to do (*as distinct from how this is to be done*). The requirements are the foundation for the system, and if they are incorrect, then the implemented system will be incorrect. *Prototyping may be employed* to assist in the definition and validation of the requirements. The process of determining the requirements, analysing and validating them, and managing them throughout the project lifecycle is termed *requirements engineering*.

The *user requirements* are determined from discussions with the customer to determine their actual needs, and they are then refined into the *system requirements*, which state the *functional* and *non-functional* requirements of the system. The specification of the user requirements needs to be unambiguous to ensure that all parties involved in the development of the system share a common understanding of what is to be developed and tested.

Requirements gathering involves meetings with the stakeholders to gather all relevant information for the proposed product. The stakeholders are interviewed, and requirements workshops are conducted to elicit the requirements from them. An early working system (prototype) is often used to identify gaps and misunderstandings between developers and users. The prototype may serve as a basis for writing the specification.

The requirements workshops are used to discuss and prioritize the requirements, as well as identifying and resolving any conflicts between them. The collected information is consolidated into a coherent set of requirements. Changes to the requirements may occur during the project, and these need to be controlled. It is essential to understand the impacts (e.g. schedule, budget, and technical) of a proposed change to the requirements prior to its approval.

*Requirements verification* is concerned with ensuring that the requirements are properly implemented (i.e. building it right) in the design and implementation. *Requirements validation* is concerned with ensuring that the right requirements are defined (building the right system), and that they are precise, complete, and reflect the actual needs of the customer.

The requirements are validated by the stakeholders to ensure that they are actually those desired, and to establish their feasibility. This may involve several reviews of the requirements until all stakeholders are ready to approve the requirements document. Other validation activities include reviews of the prototype and the design, and user acceptance testing.

The requirements for a system are generally documented in a natural language such as "English". Other notations that are employed include the visual modelling language UML (Jacobson et al. 1999) and formal specification languages such as VDM or Z for the safety-critical field.

The Agile software development methodology argues that as requirements change so quickly that a requirements document is unnecessary, such a document would be out of date as soon as it was written.

### 2.5.2 Specification of System Requirements

The specification of the system requirements of the product is essentially a statement of what the software development organization will provide to meet the business (user) requirements. That is, the detailed business requirements are a statement of what the customer wants; whereas, the specification of the system requirements is a statement of what will be delivered by the software development organization.

It is essential that the system requirements are valid with respect to the user requirements, and the stakeholders review them to ensure their validity. Traceability is employed to show that the business requirements are addressed by the system requirements.

There are two categories of system requirements, namely functional and non-functional requirements. The *functional requirements* define the functionality that is required of the system, and it may include screen shots, report layouts, or desired functionality specified as use cases. The *non-functional requirements* will generally include security, reliability, availability, performance and portability requirements, as well as usability and maintainability requirements.

### 2.5.3 Design

The design of the system consists of engineering activities to describe the architecture or structure of the system, as well as activities to describe the algorithms and functions required to implement the system requirements. It is a creative process concerned with how the system will be implemented, and its activities include architecture design, interface design, and data structure design. There are often several possible design solutions for a particular system, and the designer will need to decide on the most appropriate approach.

The design may be specified in various ways such as graphical notations that display the relationships between the components making up the design. The notation may include flow charts or various UML diagrams such as sequence diagrams, state charts, and so on. Program description languages or pseudo code may be employed to define the algorithms and data structures that are the basis for implementation.

Function-oriented design is mainly historical, and it involves starting with a high-level view of the system and refining it into a more detailed design. The system state is centralized and shared between the functions operating on that state.

Object-oriented design (OOD) is based on the concept of *information hiding* as developed by Parnas (Parnas 1972). The system is viewed as a collection of objects rather than functions, with each object managing its own state information. The system state is decentralized, and an object is a member of a class. The definition of a class includes attributes and operations on class members, and these may be inherited from super classes. Objects communicate by exchanging messages, and object-oriented design has largely replaced function-oriented design,

It is essential to verify and validate the design with respect to the system requirements, and this may be done by traceability of the design to the system requirements and design reviews to ensure that the design is fit for purpose.

### 2.5.4 Implementation

This phase is concerned with implementing the design in the target language and environment (e.g. C++ or Java) and involves writing or generating the actual code. The development team divides up the work to be done, with each programmer responsible for one or more modules. The coding activities generally include code reviews or walkthroughs to ensure that quality code is produced and to verify its correctness. The code reviews will verify that the source code conforms to the coding standards and that maintainability issues are addressed. They will also verify that the code produced is a valid implementation of the software design, and that it is fit for purpose.

Software reuse provides a way to speed up the development process. Components or objects that may be reused need to be identified and handled accordingly. The implemented code may use software components that have either being developed internally or purchased off the shelf. Open-source software has become popular in recent years, and it allows software developed by others to be used (*under an open-source licence*) in the development of applications.

The benefits of software reuse include increased productivity and a faster time to market. There are inherent risks with customized-off-the shelf (COTS) software, as the supplier may decide to no longer support the software, or there is no guarantee that the software that has worked successfully in one domain will work correctly in a different domain. It is therefore important to consider the risks as well as the benefits of software reuse and open-source software.

### 2.5.5 Software Testing

Software testing is employed to verify that the requirements have been correctly implemented, and that the software is fit for purpose, as well as identifying defects present in the software. There are various types of testing that may be conducted including *unit testing, integration testing, system testing, performance testing, and user acceptance testing*. These are described below:

**Unit Testing**
Unit testing is performed by the programmer on the completed unit (or module) prior to its integration with other modules. The programmer writes these tests, and the objective is to show that the code satisfies the design. The unit test case is generally documented, and it should include the test objectives and the expected results.

Code coverage and branch coverage metrics are often generated to give an indication of how comprehensive the unit testing has been. These provide visibility into the number of lines of code executed, as well as the branches covered during unit testing. Test tools are often employed to provide code coverage and branch coverage metrics.

The developer executes the unit tests; records the results; corrects any identified defects; and retests the software. *Test-driven development* (TDD) has become popular in the Agile world, this involves writing the unit test cases (and possibly other test cases) before the code, and the code is then written to pass the defined test cases.

**Integration Test**

The integrated system is ready to be tested once the unit testing is complete and when all of the individual units work correctly in isolation. The development team generally performs this type of testing, and the objective is to verify that all of the modules and their interfaces work correctly together. Modules that work correctly in isolation may fail when they are integrated with other modules, and the developers will analyse the cause of failure; make appropriate corrections; and retest until the integrated system works correctly.

**System Test**

The purpose of system testing is to verify that the implementation is valid with respect to the system requirements. It involves the specification and execution of system test cases to verify that the system requirements have been correctly implemented. An independent test group generally conducts this type of testing, and the system test cases are traceable to the system requirements.

Any system requirements that have been incorrectly implemented will be identified, and defects are logged and reported to the developers. The developers make the appropriate corrections, the test group verifies that the new version of the software is correct, and regression testing is conducted to verify system integrity. System testing may include security testing, usability testing, and performance testing.

The preparation of the system test environment requires detailed planning, and it may involve ordering special hardware and tools. It is important that the test environment is set up early to ensure that it is ready on time for the execution of the test cases.

**Performance Test**

The purpose of performance testing is to ensure that the performance of the system is within the bounds specified by the non-functional requirements. It may include *load performance testing*, where the system is subjected to heavy loads over a long period of time, and *stress testing*, where the system is subjected to heavy loads during a short time interval.

*Performance testing often involves the simulation of many users* using the system and involves measuring the response times for various activities. Test tools are employed to simulate a large number of users and heavy loads. This type of testing is also employed to determine if the system is scalable to support future growth.

**User Acceptance Test**

The objective of UAT testing is to demonstrate that the product satisfies the business requirements and meets the customer expectations. Upon its successful completion, the customer will be happy to accept the product. It is usually performed under controlled conditions at the customer site, and its operation will closely resemble the real-life behaviour of the system. The customer will see the product in operation and will make an informed judgment as to whether the system is fit for purpose.

### 2.5.6 Support and Maintenance

This phase continues after the release of the software product to the customer. Software systems often have a long lifetime, and the software needs to be continuously enhanced over its lifetime to meet the evolving needs of the customers. This may involve regular releases of new functionality and corrections to known defects.

Any problems that the customer identifies with the software are reported as per the customer support and maintenance agreement. The support issues will require investigation, and the issue may be *a defect in the software*, *an enhancement to the software*, or *due to a misunderstanding*. The support and maintenance team will identify the causes of any identified defects and will implement an appropriate solution to resolve. Testing is conducted to verify that the solution is correct, and that the changes made have not adversely affected other parts of the system. Mature organizations will conduct post-mortems to learn lessons from the defect[13] and will take corrective action to prevent a reoccurrence.

*The presence of a maintenance phase suggests an acceptance of the reality that problems with the software will be identified post-release*. The goal of building a correct and reliable software product the first time is very difficult to achieve, and the customer is always likely to find some issues with the released software product. It is accepted today that the quality needs to be built into each step in the development process, with the role of software inspections and testing to identify as many defects as possible prior to release, and to minimize the risk that serious defects will be found post-release.

The effective in-phase inspections of the deliverables will influence the quality of the resulting software and lead to a corresponding reduction in the number of defects. The testing group plays a key role in verifying that the system is correct, and in providing confidence that the software is fit for purpose and ready to be released. The approach to software correctness involves testing and retesting, until

[13]This is essential for serious defects that have caused significant inconvenience to customers (e.g. a major telecom outage). The software development organization will wish to learn lessons to determine what went wrong in its processes that prevented the defect from been identified during peer reviews and testing. Actions to prevent a reoccurrence will be identified and implemented.

the testing group believes that all defects have been eliminated. Dijkstra (1972) comments on testing are well-known:

> Testing a program demonstrates that it contains errors, never that it is correct.

That is, irrespective of the amount of time spent testing, it can never be said with absolute confidence that all defects have been found in the software. Testing provides increased confidence that the program is correct, and statistical techniques may be employed to give a measure of the software reliability.

Many software companies may consider one defect per thousand lines of code (KLOC) to be reasonable quality. However, if the system contains one million lines of code, this is equivalent to a thousand post-release defects, which is unacceptable.

Some mature organizations have a quality objective of three defects per million lines of code, which was introduced by Motorola as part of its six-sigma ($6\sigma$) program. It was originally applied it to its manufacturing businesses and subsequently applied to its software organizations. The goal is to reduce variability in manufacturing processes and to ensure that the processes performed within strict process control limits.

## 2.6 Software Inspections

Software inspections play an important role in building quality into the software and in reducing the cost of poor quality. There are several well-known inspection methodologies such as the Fagan Methodology (Fagan 1976) and Gilb's approach (Gilb and Graham 1994), and we briefly discussed the Fagan methodology in Chap. 1.

The seven-step Fagan process identifies and removes errors in work products. It mandates that requirement documents, design documents, source code, and test plans are all formally inspected by experts independent of the author of the deliverable. There are several *roles* defined in the process including the *moderator* who chairs the inspection. The *reader's* responsibility is to read or paraphrase the particular deliverable, and *the author* is the creator of the deliverable and has a special interest in ensuring that it is correct. The *tester* role is concerned with the testing viewpoint.

The inspection meeting will consider whether the design is correct with respect to the requirements, and whether the source code is correct with respect to the design.

## 2.7 Software Project Management

The timely delivery of quality software requires good management and engineering processes. The project management activities include the following:

- Estimation of cost, effort, and schedule for the project
- Identifying and managing risks
- Preparing the project plan
- Preparing the initial project schedule and key milestones
- Obtaining approval for the project plan and schedule
- Staffing the project
- Monitoring progress, budget, schedule, effort, risks, issues, change requests, and quality
- Taking corrective action
- Replanning and rescheduling
- Communicating progress to affected stakeholders
- Preparing status reports and presentations.

The project plan will contain or reference several other plans such as the project quality plan; the communication plan; the configuration management plan; and the test plan.

Project estimation and scheduling are difficult as often software projects are breaking new ground, and previous estimates may not be a good basis for estimation for the current project. Often, unanticipated problems can arise for technically advanced projects, and the estimates may be optimistic. Gantt charts are employed for project scheduling, and these show the work breakdown for the project, as well as task dependencies and allocation of staff to the various tasks.

The effective management of risk during a project is essential to project success. Risks arise due to uncertainty, and the risk management cycle involves[14] risk identification; risk analysis and evaluation; identifying responses to risks; selecting and planning a response to the risk; and risk monitoring. The risks are logged, and the likelihood of each risk arising and its impact is then determined. The risk is assigned an owner and an appropriate response to the risk determined.

## 2.8 CMMI Maturity Model

The CMMI is a framework to assist an organization in the implementation of best practice in software and systems engineering. It is an internationally recognized model for software process improvement and assessment and is used worldwide by thousands of organizations. It provides a solid engineering approach to the development of software, and it supports the definition of high-quality processes for the various software engineering and management activities.

It was developed by the Software Engineering Institute (SEI) who adapted the process improvement principles used in the manufacturing field to the software field. They developed the original CMM in the early 1990s and its successor the

[14]These are the risk management activities in the Prince2 methodology.

CMMI in 2001. The CMMI states *what the organization needs to do* to mature its processes rather than *how this should be done.*

The CMMI consists of five maturity levels with each maturity level consisting of several process areas. Each process area consists of a set of goals, which are implemented by practices related to that process area. Level two is focused on management practices; level three is focused on engineering and organization practices; level four is concerned with ensuring that key processes are performing within strict quantitative limits; and level five is concerned with continuous process improvement. Maturity levels may not be skipped in the staged representation of the CMMI, as each maturity level is the foundation for the next level. The CMMI and Agile are compatible, and CMMI v1.3 (released in 2010) supports Agile software development.

The CMMI allows organizations to benchmark themselves against other organizations. This is done by a formal SCAMPI appraisal conducted by an authorized lead appraiser. The results of the appraisal are generally reported back to the SEI, and there is a strict qualification process to become an *authorized lead appraiser*. An appraisal is useful in verifying that an organization has improved, and it enables the organization to prioritize improvements for the next improvement cycle. The CMMI is discussed in more detail in Chap. 16 of O'Regan (2019a).

## 2.9 Formal Methods

Dijkstra and Hoare have argued that the way to develop correct software is to derive the program from its specifications using mathematics, and to employ *mathematical proof* to demonstrate its correctness with respect to the specification. This offers a rigorous framework to develop programs adhering to the highest quality constraints. However, in practice, mathematical techniques have proved to be cumbersome to use, and they are used in specialized areas in industry.

The *safety-critical area* is one area in which mathematical techniques have been successfully applied. There is a need for extra rigour in this domain, and mathematical techniques can demonstrate the presence or absence of certain desirable or undesirable properties (e.g. "*when a train is in a level crossing, then the gate is closed*").

Spivey (1992) defines a "*formal specification*" as the use of mathematical notation to describe in a precise way the properties which an information system must have, without unduly constraining the way in which these properties are achieved. It describes *what* the system must do, as distinct from *how* it is to be done. This abstraction away from implementation enables questions about what the system does to be answered, independently of the detailed code. Further, the unambiguous nature of mathematical notation avoids the problem of ambiguity in an imprecisely worded natural language description of a system.

The formal specification thus becomes the key reference point for the different parties concerned with the construction of the system and is a useful way of promoting a common understanding for all those concerned with the system. The term "*formal methods*" is used to describe a formal specification language and a method for the design and implementation of computer systems.

The specification is written precisely in a mathematical language. The derivation of an implementation from the specification may be achieved via *step-wise refinement*. Each refinement step makes the specification more concrete and closer to the actual implementation. There is an associated *proof obligation* that the refinement be valid, and that the concrete state preserves the properties of the more abstract state. Thus, assuming the original specification is correct and the proofs of correctness of each refinement step are valid, then there is a very high degree of confidence in the correctness of the implemented software.

Formal methods have been applied to a diverse range of applications, including circuit design, artificial intelligence, specification of standards, specification, and verification of programs, etc. They are discussed in more detail in Chap. 13.

## 2.10 Review Questions

1. Discuss the research results of the Standish Group the current state of IT project delivery?
2. What are the main challenges in software engineering?
3. Describe various software lifecycles such as the waterfall model and the spiral model.
4. Discuss the benefits of Agile over conventional approaches. List any risks and disadvantages?
5. Describe the purpose of the CMMI? What are the benefits?
6. Describe the main activities in software inspections.
7. Describe the main activities in software testing.
8. Describe the main activities in project management?
9. What are the advantages and disadvantages of formal methods?

## 2.11 Summary

The birth of software engineering at the NATO conference in 1968 in Germany highlighted the problems that existed in the software sector in the late 1960s, and the term "*software crisis*" was coined to refer to these. The conference led to the realization that programming is quite distinct from science to mathematics, and that software engineers need to be properly trained to enable them to build high-quality products that are safe for the public.

Programmers are like engineers in the sense that they build products. Therefore, programmers need to receive an appropriate education in engineering as part of their training. The education of traditional engineers includes training on product design, and an appropriate level of mathematics.

Software engineering is a systematic approach to the development and maintenance of the software, and it requires a precise statement of the requirements of the software product, and then the design and development of a solution to meet these requirements. It includes methodologies to design, develop, implement, and test software as well as sound project management, quality management, and configuration management practices. Support and maintenance of the software need to be properly addressed.

Software process maturity models such as the CMMI have become popular in recent years. They place an emphasis on understanding and improving the software process to enable software engineers to be more effective in their work.

## References


Beck K (2000) Extreme programming explained. Embrace change. Addison Wesley, Boston

Boehm B (1988) A spiral model for software development and enhancement. Computer

Brooks F (1975) The mythical man month. Addison Wesley, Boston

Brooks F (1986) No silver bullet. Essence and accidents of software engineering. Information processing. Elsevier, Amsterdam

Dijkstra EW (1972) Structured programming. Academic Press, Cambridge

Fagan M (1976) Design and code inspections to reduce errors in software development. IBM Syst J 15(3)

Gilb T, Graham D (1994) Software inspections. Addison Wesley, Boston

Jacobson I, Booch G, Rumbaugh J (1999) The unified software modelling language user guide. Addison-Wesley, Boston

Naur P, Randell B (1969) Software engineering: report on a conference sponsored by the NATO Science Committee, Garmisch, Germany, 7th to 11th October 1968. Scientific Affairs Division, NATO, Brussels

Naur P, Randell B (1975) Software engineering. Petrocelli, IN, Buxton. Report on two NATO conferences held in Garmisch, Germany (October1968) and Rome, Italy (October 1969)

Office of Government Commerce (2004) Managing successful projects with PRINCE2. Office of Government Commerce, UK

O'Regan G (2006) Mathematical approaches to software quality. Springer, London

O'Regan G (2010) Introduction to software process improvement. Springer, London

O'Regan G (2014) Introduction to software quality. Springer, London

O'Regan G (2017a) Concise guide to software engineering. Springer, Berlin

O'Regan G (2017b) Concise guide to formal methods. Springer, Berlin

Parnas D (1972) On the criteria to be used in decomposing systems into modules. Commun ACM 15(12)

Royce W (1970) The software lifecycle model (waterfall model). In: Proceedings of WESTCON, August, 1970

Rumbaugh J et al (1999) The unified software development process. Addison Wesley, Boston

Spivey JM (1992) The Z notation. A reference manual. Prentice Hall International Series in Computer Science

# 3 Fundamentals of Software Testing

**Key Topics**

Test planning
Test case design
Unit testing
System testing
Performance testing
Psychology of software tester
Acceptance testing
White box testing
Black box testing
Test tools
Test environment
Test reporting

## 3.1 Introduction

Testing plays a key role in verifying the correctness of software and confirming that the requirements have been correctly implemented. It is a constructive and destructive activity in that while, on the one hand, it aims to verify the correctness of the software, on the other hand, it aims to find as many defects as possible in the software. The vast majority of defects (e.g. 80%) are detected by software inspections in a mature software organization, with the remainder detected by the


G. O'Regan, *Concise Guide to Software Testing*,
Undergraduate Topics in Computer Science,
https://doi.org/10.1007/978-3-030-28494-7_3

various types of testing carried out during the project. Testing has been defined in TMaps (2004) as:

> Testing is a process of planning, preparing, executing, and analysing, aimed at establishing the characteristics of an information system and demonstrating the differences between the actual status and the required status.

Software testing involves defining the test conditions and designing the test cases and then executing the test cases. This is followed by analysis and reporting of the results. Software testing provides confidence that the product is ready for release to potential customers, and the recommendation of the testing department plays a key role in the decision on whether to release the software product or not. The test manager highlights any risks associated with the product, and these are carefully considered to ensure that they can be managed on release. The test manager and test department are influential in an organization by providing strategic advice on product quality and in encouraging organization change to improve the quality of the software product through the use of best practice in software and system engineering.

The testers need a detailed understanding of the software requirements to develop appropriate test cases to verify the correctness of the software. Test planning commences at the early stages of the project, and testers play a role in building quality into the software product through software inspections. The testers will generally participate in the review of the requirements, as the testing viewpoint is important in ensuring that the requirements are correct and testable.

The test plan for the project is documented (this could be part of the project plan or a separate document), and it includes the scope of the testing, the personnel involved, the resources and effort required, the key milestones, the definition of the test environment, any special hardware and test tools required, and the planned test schedule. There is a separate test specification plan for the various types of testing, which records the test cases, including the purpose of each test case, the inputs and expected outputs, and the test procedure for the execution of the particular test case.

Several types of testing are performed during the project, including unit, integration, system, regression, performance, and user acceptance testing. The software developers perform the unit testing, and the objective is to verify the correctness of a module. This type of testing is termed *"white box"* testing and is based on knowledge of the internals of the software module. White box testing typically involves checking that every path in a module has been tested, and it involves defining and executing test cases to ensure code and branch coverage. The objective of *"black box"* testing is to verify the functionality of a module (or feature or the complete system itself), and knowledge of the internals of the software module is not required.

Test reporting ensures that all project participants understand the current quality of the software, as well as understanding what needs to be done to ensure that the product achieves the desired quality criteria. The test status is reported regularly during the project, and once the tester discovers a defect, a problem report is opened, and the problem is analysed and corrected by the software developers. The problem may indicate a genuine defect, a misunderstanding by the tester, or a request for an enhancement.

An *independent test group* is more effective than a test group that is directly reporting to the development manager. The independence of the test group helps ensuring that quality is not compromised when the project is under pressure to make its committed delivery dates. A good test group will play a proactive role in quality improvement, and this may involve participation in the analysis of the defects identified during testing at the end of the project, with the goal of prevention or minimization of the reoccurrence of the defects.

Real-world issues such as the late delivery of the software from the developers often complicate software testing. Software development is challenging and deadline-driven, and missed developer deadlines may lead to compression of the testing schedule, as the project manager may wish to stay with the original project schedule due to commitments to the stakeholders. There are risks associated with shortening the test cycle, as the testers may be unable to complete the planned test activities. This means that there may be insufficient data to make an informed judgment as to whether the software is ready for release, leading to risks that a defect-laden product may be shipped to the customer.

Test departments may be understaffed, as management may consider additional testers to be expensive. The test manager needs to be assertive in presenting the test status of the project and in clearly communicating the quality and associated risks. The recommendation of the test manager needs to be carefully considered by the project manager and other stakeholders prior to the release of the software.

## 3.2 Software Test Process

The quality of the testing is dependent on the maturity of the test process, and a good test process will include test planning, test case analysis and design, test execution, and test reporting. A simplified test process is sketched in Fig. 3.1, and it will typically include:

- Test planning and risk management
- Dedicated test environment and test tools
- Test case definition
- Test automation
- Test execution
- Formality in handover to test department
- Test result analysis
- Test reporting
- Measurements of test effectiveness
- Test management
- Lessons learned and test process improvement.

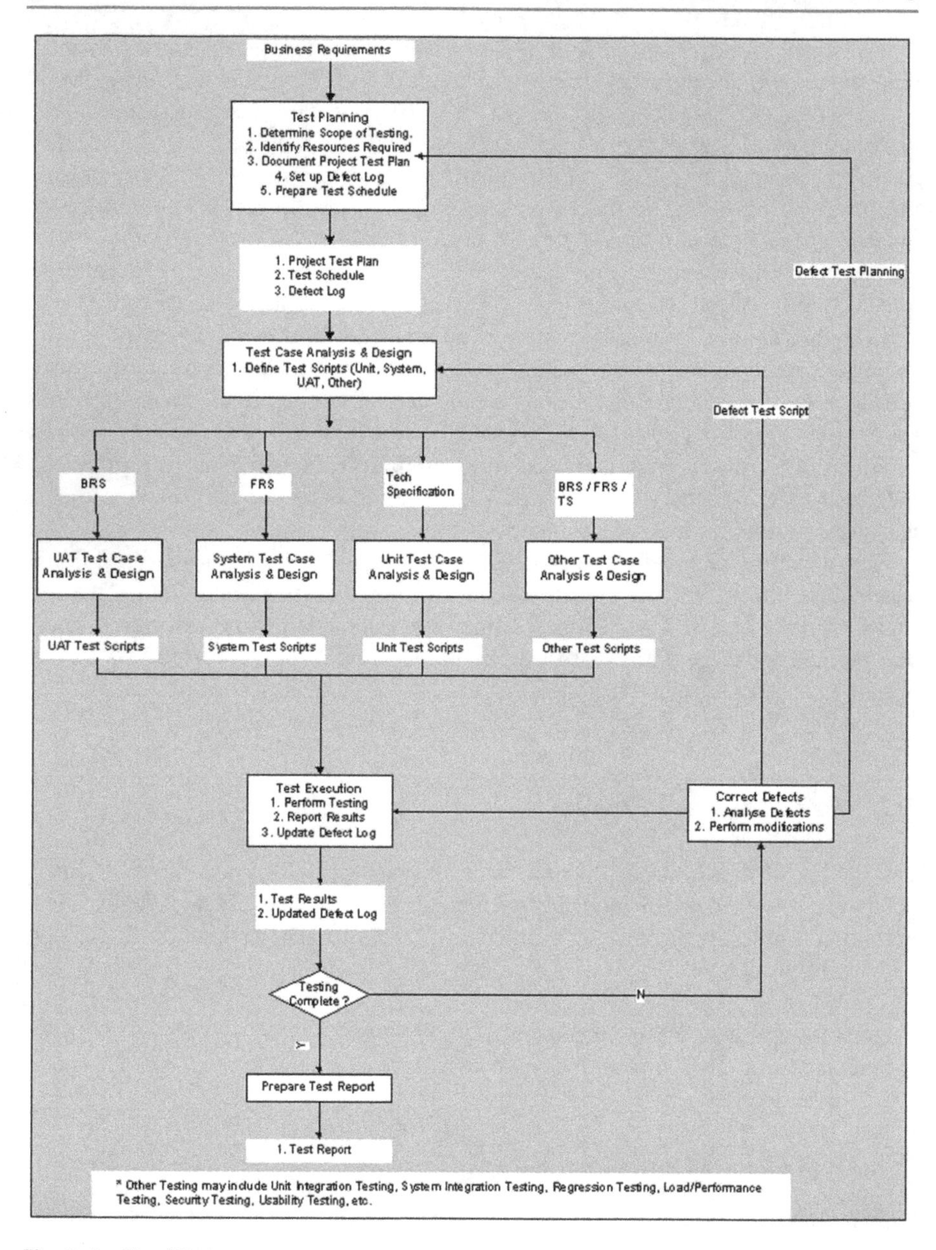


**Fig. 3.1** Simplified test process

Test planning consists of a documented plan defining the scope of testing and the various types of testing to be performed, the definition of the test environment, the required hardware or software for the test environment, the estimation of effort and resources for the various activities, risk management, the deliverables to be produced, the key test milestones, and the test schedule.

The test plan is reviewed to ensure its fitness for purpose and to obtain commitment to the plan, as well as ensuring that all involved understand and agree to their responsibilities. The test plan may be revised in a controlled manner during the project. It is described in more detail in Sect. 3.3.

The test environment varies according to the project requirements and business environment. Large organizations may employ dedicated test laboratories, whereas a single workstation may be sufficient for a small organization. A dedicated test environment may require significant capital investment, but it will pay for itself by identifying defects and verifying that the software is fit for purpose, leading to reductions in the cost of poor quality.

The test environment includes the hardware and software needed to verify the correctness of the software. It is generally defined early in the project to allow any specialized hardware or software to be ordered in time. It may include simulation tools, automated regression and performance test tools, as well as tools for defect tracking and reporting.

The software developers produce a software build under configuration management control, and the build is verified for integrity to ensure that testing may commence. There is generally a formal or informal handover of the software to the test department, which includes criteria that must be satisfied for testing to commence. The test department must be ready to carry out the testing, with the test cases defined and the test environment set-up.

Several types of testing employed to verify the correctness of the software are described in Table 3.1.

The effectiveness of the testing is dependent on the definition of good test cases, which need to be complete in the sense that their successful execution will provide confidence in the correctness of the software. Hence, the test cases must cover the software requirements, and a traceability matrix may be employed to map the requirements to the design and test cases. The traceability matrix provides confidence that each requirement has a corresponding test case for verification. The test cases will include the purpose of the test case, the set-up required, inputs, the procedure, and expected outputs.

The test execution will follow the procedure defined in the test cases, and the tester will compare the actual results obtained with the expected results. The test completion status will be passed, failed, or blocked (if unable to run at this time). The test results summary will indicate which test cases could be executed, which passed, which failed, and which test cases could not be executed.

The test results are documented including detailed information on the passed and failed tests. This will assist the software developers in identifying the precise causes of failure and determining an appropriate solution. The developers and tester will agree to open a defect report in the defect tracking system to track the correction of each defect.

*Test levels* refer to a group of testing activities that are organized and managed together. A test level is linked to the responsibilities in a project (Table 3.2).

**Table 3.1** Types of testing

| Test type | Description |
|---|---|
| Unit testing | This testing is performed by the software developers to verify the correctness of the software modules |
| Component testing | This testing is performed by the software developers to verify the correctness of the software components, i.e. to ensure that the component is correct and may be reused |
| System testing | This testing is generally carried out by an independent test group to verify the correctness of the complete system |
| Performance testing | This testing is generally carried out by an independent test group to ensure that the performance of the system is within the defined parameters. It may require tools to simulate clients and heavy loads, and precise measurements of performance are made |
| Load/stress testing | This testing is used to verify that the system performance is within the defined limits for heavy system loads over long or short periods of time |
| Browser compatibility | This testing is specific to Web-based applications and verifies that the website functions correctly with the supported browsers |
| Usability testing | This testing verifies that the software is easy to use and that the look and feel of the application is good |
| Security testing | This testing verifies that the confidentiality, integrity, and availability requirements are satisfied |
| Regression testing | This testing verifies that the core functionality is preserved following changes or corrections to the software. Test automation may be employed to increase its productivity and efficiency |
| Test simulation | This testing simulates part of the system where the real system currently does not exist, or where the real-life situation is hard to replicate |
| Acceptance testing | This testing carried out by the customer to verify that the software is fit for purpose and matches the customer's expectations |

**Table 3.2** Test levels

| Test level | Description |
|---|---|
| Component testing | Each component is tested separately prior to integration with others. It may include functional, non-functional, and structural tests. The test cases are based on software design and code structure |
| Integration testing | The integrated components are tested together with functional, non-functional, and structural tests |
| System testing | The integrated system is tested by a dedicated test team using functional and non-functional tests (sometimes structural testing—e.g. page navigation) |
| Acceptance testing | This testing is the responsibility of the customer, and the goal is to verify that the software is fit for purpose and matches the customer's expectations. It involves user acceptance testing and operational acceptance testing. It may involve contractual and compliance acceptance testing and alpha/beta testing |

The test status (Fig. 9.11) consists of the number of tests planned, the number of test cases run, the number that have passed, and the number of failed and blocked tests. The test status is reported regularly to management during the testing cycle. The test status and test results are analysed and extra resources provided where necessary to ensure that the product is of high quality with all agreed defects corrected prior to the acceptance of the product.

Test tools and test automation are used to support the test process and lead to improvements in quality, reduced cycle time, and increased productivity. Tool selection needs to be performed in a controlled manner, and it is best to identify the requirements for the tool first and then to examine a selection of tools to determine which best meets the requirements. Tools may be applied to test management and reporting and to the various types of testing.

A good test process will maintain measurements to determine its effectiveness, and an end of testing review is conducted to identify any lessons that need to be learned for continuous improvement. The test metrics employed will answer questions such as:

- What is the current quality of the software?
- How stable is the product at this time?
- Is the product ready to be released at this time?
- How good was the quality of the software that was handed over?
- How does the product quality compare to other products?
- How much testing remains to be done?
- How effective was the testing performed on the software?
- How many open problems are there and how serious are they?
- What are the key risks and are they all managed?

## 3.3 Software Test Planning and Scheduling

Testing is a subproject of a project and needs to be managed as such, and so good planning and monitoring and control are required. The IEEE 829 standard includes a template for test planning, which involves defining the scope of the testing to be performed; defining the test environment; estimating the effort required to define the test cases and to perform the testing; identifying the resources needed (including people, hardware, software, and tools); assigning people to the tasks; defining the schedule; and identifying any risks to the testing and managing them.

The tracking of the testing involves monitoring progress and taking corrective action to ensure quality and schedule are achieved; replanning when the scope of the testing has changed; preparing test reports to give visibility to management (including the number of tests planned, executed, passed, blocked, and failed); retesting corrections to the failed or blocked tests; managing risks; and providing a final test report with a recommendation to go to acceptance testing.

**Table 3.3** Simple test schedule

| Activity | Resource name(s) | Start date | End/replan date | Comments |
|---|---|---|---|---|
| Review requirements | Test Team | 15.02.2019 | 16.02.2019 | Complete |
| Project test plan and review | J. DiNatale | 15.02.2019 | 28.02.2019 | Complete |
| System test plan/review | P. Cuitino | 01.03.2019 | 22.03.2019 | Complete |
| Performance test plan/review | L. Padilla | 15.03.2019 | 31.03.2019 | Complete |
| Regression plan/review | X. Yun | 01.03.2019 | 15.03.2019 | Complete |
| Set-up test environment | X. Yun | 15.03.2019 | 31.03.2019 | Complete |
| System testing | P. Cuitino | 01.04.2019 | 31.05.2019 | In progress |
| Performance testing | L. Padilla | 15.04.2019 | 07.05.2019 | In progress |
| Regression testing | L. Padilla | 07.05.2019 | 31.05.2019 | In progress |
| Test reporting | J. DiNatale | 01.04.2019 | 31.05.2019 | In progress |

Table 3.3 presents a simple test schedule for a small project, and the test manager will usually employ Microsoft Project for scheduling and tracking of larger projects (e.g. Fig. 5.2). The activities in the schedule are tracked and updated to record progress, and dates are revised as appropriate. The project manager will track the key test milestones and will maintain close contact with the testing manager.

It is prudent to consider risk management early in test planning, to identify risks that could potentially arise during the testing, and to manage them accordingly. The probability of occurrence of each risk and its impact is determined.

## 3.4 Test Case Design and Definition

Several types of testing that may be performed during the project were described in Table 3.1, and there may be a separate test plan for unit, system, and UAT testing. The unit tests are based on the software design; the system tests are based on the system requirements; and the UAT tests are based on the business (or user) requirements.

Each of these test plans contains test scripts (e.g. the unit test plan contains the unit test scripts and so on), and the test scripts are traceable to the design (for the unit tests), and for the system requirements (for the system test scripts). The unit tests are more focused on white box testing, whereas the system test and UAT tests are focused on black box testing. A test script generally includes:

– Test case ID
– Test type (e.g. unit, system, UAT)
– Objective/description
– Test script steps
– Expected results
– Actual results
– Tested by.

## 3.5 Test Execution

The software developers will carry out the unit and integration testing as part of the normal software development activities. They will correct any identified defects, and the development continues until all unit and integration tests pass, and the software is fit to be released to the test group.

The test group will usually be *independent* (i.e., it has an independent reporting channel from the development manager), and the test activities will usually include system testing, performance testing, usability testing, and so on. There is usually a formal handover from development to the test group prior to the commencement of testing, and the handover criteria need to be satisfied in order for the software to be accepted for testing by the test group.

Test execution then commences and the testers run the system tests and other tests, log any defects in the defect tracking tool, and communicate progress to the test manager. The test status is communicated to management, and the developers correct the identified defects and produce new releases. The test group retests the failed and blocked tests and performs regression testing to ensure that the core functionality remains in place. This continues until the quality goals for the project have been achieved.

## 3.6 Test Reporting and Project Sign-off

The test manager will report progress regularly during the project. The report provides the current status of testing for the project including:

- Quality status (including tests run, passed, and blocked).
- Risks and issues
- Status of test schedule
- Deliverables planned (next period).

The test manager discusses the test status with management and highlights the key risks and issues to be dealt with. The test manager may require management support to deal with these.

The test status is important in judging whether the software is ready to be released to the customer. Various quality metrics may be employed to measure the quality of the software, and the key risks and issues are considered. The test manager will make a recommendation to release or not based on the actual test status. One useful metric (one of many to consider) is the cumulative arrival rate (Fig. 9.12) that gives an indication of the stability of the product.

The slope of the curve is initially steep as testing commences and defects are detected. As testing continues and defects are corrected and retested, the slope of the curves levels off, and overtime the indications are that the software has stabilized, and is potentially ready to be released to the customer.

However, it is important not to rush to conclusions based on an individual measurement. For example, Fig. 9.12 indicates that testing halted on 13 May with no testing since then and that would explain why the defect arrival rate per week is zero. Careful investigation needs to be done before the interpretation of a measurement is made, and often several measurements rather than one are employed to make a sound decision.

## 3.7 Testing and Quality

Testing allows the quality of the software to be measured in terms of the defects found. It provides confidence in the quality of the software, and a properly designed test case that passes reduces the level of risk with the system. Further, the quality of the system generally increases (reliability growth) once defects identified during testing have been corrected and verified.

The recommendation of the test manager is carefully considered in the decision on whether to release the software product or not. Decision-making is based on objective facts, and measurements are generally employed to assess the quality of the software.

The open-problem status (Figs. 9.7 and 9.9), the problem arrival rate (Fig. 9.12), and the cumulative problem arrivals' rate (Fig. 9.13) give an indication of the quality and stability of the software product and may be used with other measures to decide on whether it is appropriate to release the software or whether further testing should be done.

### 3.7.1 What Is a Software Defect?

A *defect* is a flaw in the software that causes the software to fail to perform its required function. A defect that is encountered during program execution leads to a software failure.

A defect arises due to a developer making an error that produces a defect in the code. The code where the defect is present may be executed, which results in the software failing to do what it is required (i.e. failure). Not all defects result in failure as some defects may be on rarely used execution paths.

The defect density of the software is the number of defects in the software divided by the number of lines of code.

### 3.7.2 Is Exhaustive Testing Possible?

It may seem like a reasonable approach to perform exhaustive testing of the software for black box and white box testing. However, exhaustive black box testing would involve using every possible test input condition (valid and invalid), and the

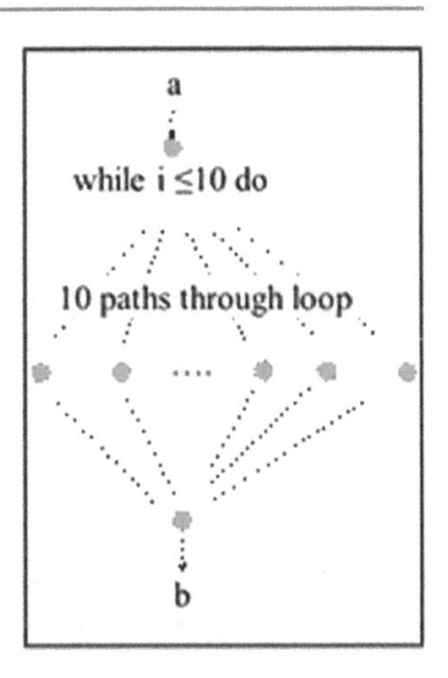


**Fig. 3.2** Number of paths through a trivial program

number of test cases rapidly becomes astronomical (potentially infinite). Similarly, it is unrealistic in white box testing to perform exhaustive path testing and to execute all possible paths through the program. This is since the number of paths through a relatively simple program rapidly becomes astronomical and so exhaustive path coverage is unrealistic.

For the simple program in Fig. 3.2, we have a while loop that executes 10 times, and there are 10 possible paths through each iteration of the loop. This means that there are a total of $10^{10}$ possible paths (10 billion paths) through this simple program. Suppose the tester writes a test in one second, then it would take one tester over 300 years to write $10^{10}$ tests to check each possible path.

Clearly, exhaustive testing is not feasible and instead alternative approaches such as risk-based approaches to testing need to be employed.

### 3.7.3 How Much Testing Should Be Done?

The amount of software testing to be done depends on the level of business and technical risks, as well as the project risks and the constraints on time and budget. Testing should provide sufficient information to allow the stakeholders to make informed decisions as to whether further testing should be done or whether the product is ready to be released.

### 3.7.4 Testing and Quality Improvement

Test defects are valuable in the sense that they provide valuable information that allows the organization the opportunity to improve its software development process to *prevent the defects from reoccurring in the future*. A mature development organization will perform internal reviews of requirements, design, and code prior to testing. The effectiveness of the internal review process and the test process may be seen in the phase containment metric (PCE), which is discussed in Chap. 9.

Figure 9.10 indicates that the project had a phase containment effectiveness of approximately 54%. That is, the developers identified 54% of the defects (in software inspections and unit/integration testing), the system testing phase identified approximately 23% of the defects, acceptance testing identified approximately 14% of the defects, and the customer identified approximately 9% of the defects. Many organizations set goals with respect to the phase containment effectiveness of their software. For example, a mature organization might aim for their software development department to have a phase containment effectiveness goal of 80%. This means that 80% of the defects should be found by software inspections.

The improvement trends in phase containment effectiveness may be tracked over time. There is no point in setting a goal for a particular group or area unless there is a clear mechanism to achieve the goal. Thus to achieve a goal of 80% phase containment effectiveness, the organization will need to implement a formal software inspection methodology as described in Chap. 4. Training on inspections will be required, and the effectiveness of software inspections monitored and improved.

A mature organization will aim to have 0% of defects reported by the customer, and this goal requires improvements in its software inspection methodology and its software testing methodology. For example, the test process may be improved by using more effective test tools for various areas of testing (e.g. performance testing). Measurements provide a way to verify that the improvements have been successful. Each defect is potentially valuable as it, in effect, enables the organization to identify weaknesses in the software process and to target improvements.

Escaped customer defects offer an opportunity to improve the testing process, as it indicates a weakness in the test process in detecting the defect earlier in the process. The defects are categorized, causal analysis is performed, and corrective actions are identified to improve the testing process. This helps to prevent a reoccurrence of the defects, and so software testing plays an important role in quality improvement.

## 3.8 Psychology of Software Tester

The mindset for reviewing and testing is quite different from that of analysing and developing software (Fig. 3.3). There are several skills required to be an effective software tester (e.g. good attention to detail, destructive creativity), and the mindset of the tester influences the outcome of the testing. For example, if the tester is also the developer of the software, then the tester's objectives tend to be focused on demonstrating the correctness of the software, rather than in seeking defects in the software.

The effectiveness of the testing is influenced by the degree of independence of the testing, as professional testers are specialists in finding defects in the software. There are several degrees of independence in testing varying from low level of

**Fig. 3.3** Psychology of software tester. Public domain

independence where the author of the code tests to complete independence where the testing is outsourced to an external organization.

– Tests developed by the person who wrote the software
– Tests designed by another person in the same development group
– Tests designed by a person from a dedicated test group
– Tests developed by an external test organization.

Myers observed that the psychology of the person carrying out the testing influences the testing (Myers 1979). For example, if the tester believes that the purpose of the testing is to demonstrate that the software works correctly (*constructive mindset*), then the tester is likely to focus on proving this point and using inputs for which correct results will be obtained. Similarly, if the tester believes that the purpose of testing is to show that the software does not work (*destructive mindset*), then this approach helps in identifying most of the defects in the software.

However, in practice it is not possible to find all defects in commercial software, and a risk-based approach is often employed. Finally, if the tester believes that the purpose of testing is to detect as many defects as possible and to minimize the risks associated with the release of the software, then this approach is often optimal in achieving good results.

The software tester and developer share a common goal in that they both desire high-quality software, and so they need to collaborate closely to achieve the best possible outcome. The identification of defects during testing may be perceived as a criticism of the product and developers, and so it is essential that defects are communicated in a constructive and professional way. Otherwise, there is a danger that there could be bad feelings between the testers and developers, and so the relationship needs to be professional with the tester clearly communicating the known defects with the software (as well as the steps to reproduce each defect) and the developer investigating and correcting the defects.

The decision to release the software may be based on several factors such as internal measurements of its quality, business factors, time constraints, and so on. However, it is essential that the risks with the software are known and that they can be managed effectively. Software testing is a means of reducing risk in software engineering.

## 3.9 Test-Driven Development

Test-driven development (TDD) ensures that there is an emphasis on testability of the code from the earliest part of the development. The approach is to write the test cases early, and the software code is then written to pass the test cases. It is a paradigm shift from traditional software engineering, where traditionally unit tests are written and executed after the code has been written.

The test-driven development of a new feature begins with writing a suite of test cases based on the requirements for the feature, and the code for the feature is then written to pass the test cases. Initially, all tests fail as no code has been written, and so the first step is to write some code that enables the new test cases to pass. The next step is to ensure that the new feature works with the existing features, and this involves executing all new and existing test cases.

This may involve modification of the source code to enable all of the tests to pass and to ensure that all features work correctly together. The final step is improving the code without changing the functionality and restructuring the code where appropriate to improve its efficiency. The test cases are rerun to ensure that the functionality is not altered in any way. The process repeats with the addition of each new feature. TDD is described in more detail in Chap. 12.

## 3.10 E-Commerce Testing

There has been an explosive growth in electronic commerce, and website quality and performance are a key concern. A website is a software application and so standard software engineering principles are employed to verify its quality. E-commerce applications are characterized by:

- Distributed system with millions of servers and billions of participants
- High availability requirements (24 * 7 * 365)
- Look and feel of the website is highly important

- Browsers may be unknown
- Performance may be un-predictable
- Users may be unknown
- Security threats may be from anywhere
- Rapidly changing technologies.

Often a rapid application development model such as RAD/JAD or Agile is employed to design a little, implement a little, and test a little, and the standard waterfall lifecycle model is rarely employed for the front end of a Web application and. The use of lightweight development methodologies does not mean that anything goes in software development, and similar project documentation should be produced (except that the chronological sequence of delivery of the documentation is more flexible). Joint application development allows early user feedback to be received on the look and feel and correctness of the application, and the method of design a little, implement a little, and test a little is generally used for Web development. The various types of Web testing include:

- Static testing
- Unit testing
- Functional testing
- Browser compatibility testing
- Usability testing
- Security testing
- Load/performance/stress testing
- Availability testing
- Post-deployment testing

Static testing generally involves inspections and reviews of documentation. The purpose of static testing of websites is to check the content of the Web pages for accuracy, consistency, correctness, and usability and also to identify any syntax errors or anomalies in the HTML. There are tools available (e.g. NetMechanic) for statically checking the HTML for syntax correctness.

The purpose of unit testing is to verify that the content of the Web pages corresponds to the design, that the content is correct, that all the links are valid, and that the Web navigation operates correctly.

The purpose of functional testing is to verify that the functional requirements are satisfied. It may be quite complex as e-commerce applications may involve product catalogue searches, order processing, credit checking and payment processing, and the application may liaise with legacy systems. Also, testing of cookies, whether enabled or disabled, needs to be considered.

The purpose of browser compatibility testing is to verify that the Web browsers that are to be supported are actually supported. The purpose of usability testing is to verify that the look and feel of the application is good and that Web performance (loading Web pages, graphics, etc.) is good. There are automated browsing tools which go through all of the links on a page, attempt to load each link, and produce a report including the timing for loading an object or page. Usability needs to be considered early in design and is important in GUI applications.

The purpose of security testing is to ensure that the website is secure. The purpose of load, performance, and stress testing is to ensure that the performance of the system is within the defined parameters.

The purpose of post-deployment testing is to ensure that website performance remains good following deployment at the customer site, and this may be done as part of a service-level agreement (SLA). A SLA typically includes a penalty clause if the availability of the system or its performance falls outside the defined parameters. Consequently, it is important to identify performance and availability issues early before they become a problem. Thus, post-deployment testing includes monitoring of website availability, performance, and security, and taking corrective action. E-commerce sites operate 24 h a day for 365 days a year, and major financial loss could potentially be incurred in the case of a major outage.

## 3.11 Traceability of Requirements

The objective of requirements traceability is to verify that all of the requirements have been implemented and tested. One way to do this would be to examine each requirement number and to go through every part of the design document to find any reference to the particular requirement number, and similarly to go through the test plan and find any reference to the requirement number. This would demonstrate that the particular requirement number has been implemented and tested.

A more effective mechanism to do this is with a traceability matrix (Table 6.5). This may be a separate document or part of the test documents. The idea is that a mapping between the requirement numbers and the associated test cases is defined, and this provides confidence that all of the requirements have been implemented and tested. A traceability matrix provides confidence that each requirement number has been implemented in the software design and tested via the test plan. Requirement traceability is discussed in more detail in Sect. 6.5.

## 3.12 Software Maintenance and Evolution

Software maintenance is the process of changing a system after it has been delivered to the customer, and it involves correcting any defects that are present in the software and enhancing the system to meet the evolving needs of the customer. The defects may be due to coding, design, or requirements' errors, with coding defects less expensive to fix than requirements' defects. The resolution to the defects involves identifying the affected software components and modifying them and verifying that the solution is correct and that no new problems have been introduced.

Software systems often have a long lifetime (e.g., some systems have a lifetime of 20-30 years), and so the software needs to be continuously enhanced over its lifetime to meet the evolving needs of the customer. Software evolution is concerned with the continued development and maintenance of the software after its initial release, with new releases of the software prepared each year. Each new release includes new functionality and corrections to the known defects.

Maintenance testing plays a key role in verifying that the new release is fit for purpose, and the testing performed depends on the changes made to the system and the associated risks.

## 3.13 Software Test Tools

Test tools are employed to support the test process and to enhance quality and increase productivity. Tool selection needs to be planned, and the selection of a particular tool involves defining the requirements of the proposed tool and identifying candidate tools to evaluate against the requirements. Each tool is assessed to yield an evaluation profile, and the results analysed to enable an appropriate choice to be made.

There are various tools to support testing such as test planning and management tools; defect tracking tools; regression test automation tools; performance tools; and so on (Fig. 3.4). There are tools available from various vendors such as Compuware, Software Research, Inc., HP, LDRA, McCabe and Associates, and IBM Rational.

*Test Management Tools*

There are various test management tools available and their main features are:

- Management of entire testing process
- Test planning
- Test status and reporting
- Graphs for presentation
- Defect tracking system
- Support for many testers
- Audit trail proof that testing has been done
- Test automation
- Support for various types of testing.

*Miscellaneous Testing Tools*
There is a wide collection of test tools to support activities such as static testing, unit testing, system testing, performance testing, and regression testing. Code coverage tools are useful for unit testing, and they analyse source code files to

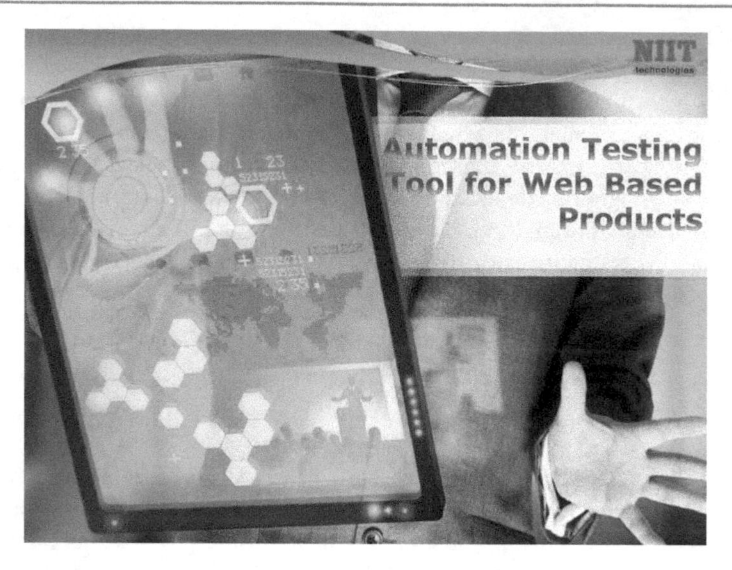


**Fig. 3.4** Automated testing tools. Creative Commons

report on areas of code that were not executed at run time, thereby facilitating the identification of missing test data. They generally provide visual reports of the code areas that were executed.

Regression testing involves rerunning existing test cases to verify that the software remains correct following changes made to correct defects or implement new technology. It is often automated with capture and playback tools, and these tools capture, verify, and replay user interactions and allow regression testing to be automated.

The purpose of performance testing to verify that system performance is within the defined limits, and it requires measures on the server side, network side, and client side (e.g. processor speed, disc space used, memory used). Performance testing tools allow the software application to be tested with hundreds or thousands of concurrent users to determine its performance under heavy loads. It allows the scalability of the software system to be tested, to determine if can support predicted growth.

The decision on whether to automate and what to automate often involves a test process improvement team. It tends to be difficult for a small organization to make a major investment in test tools (especially if the projects are small). However, larger organizations will require a more sophisticated testing process to ensure that high-quality software is consistently produced. Tools to support software testing are discussed in Chap. 10.

## 3.14 Review Questions

1. Describe the main activities in test planning.
2. What does the test environment consist of? When should it be set up?
3. Explain the traceability of the requirements to the test cases?
4. Describe the various types of testing that may be performed.
5. Investigate available test tools to support testing? What are their benefits?
6. Describe an effective way to evaluate and select a test tool.
7. What characteristics make e-commerce testing unique from other domains.
8. Discuss the influence of the test manager.
9. Explain test-driven development.

## 3.15 Summary

This chapter discussed the fundamentals of software testing and how testing is used to verify that the software is of high quality and fit to be released to customers. Testing is both a constructive and destructive activity, in that while, on the one hand, it aims to verify the correctness of the software, on the other hand, it aims to find as many defects as possible.

Several test activities were discussed including test planning, setting up the test environment, test case definition, test execution, defect reporting, and test management and reporting. We discussed black and white box testing, unit and integration testing, system testing, performance testing, security, and usability testing. Testing in an e-commerce environment was considered.

The mindset of the software tester is important where a destructive mindset helps in detecting as many defects as possible, whereas a constructive mindset is often focused on confirming correctness rather than finding defects. Often a mixture of a destructive mindset and destructive helps in minimizing the risks associated with the release of the software.

Test reporting enables all project participants to understand the current quality of the software and to understand what needs to be done to ensure that the product meets the required quality criteria.

We discussed tools to support the testing process, and tool selection and evaluation should be done formally. Metrics are useful in providing visibility into test progress and into the quality of the software. The role of testing in promoting quality improvement was discussed.

Testing is often complicated by the late delivery of the software from the developers, and this may lead to the compression of the testing schedule. The recommendation of the test manager on whether to release the product needs to be carefully considered.

## References

Myers G (1979) The art of software testing. Wiley, Hoboken
Tmaps (2004) TMap home pages. Sogeti Nederland B.V., Amersfoort. http://www.tmap.net

# Static Testing

# 4

**Key Topics**

Informal reviews
Structured walk-through
Fagan inspection
Gilb inspections
Economic benefits of inspections
Inspection guides
Entry and exit criteria
Automated software inspections

## 4.1 Introduction

*Static testing* (as distinct from *dynamic testing*) is a form of software testing that involves a systematic examination of the software code and documentation without execution of the code. It may be conducted manually or through the use of specialized software testing tools. There are several types of static testing such as code analysis, code reviews, structured walk-throughs, informal reviews and software inspections.

The objective of software inspections is to build quality into the software product, rather than adding quality later. There is clear evidence that the cost of correction of a defect increases later that it is detected, and it is therefore more cost effective to build quality in rather than adding it later in the development cycle. Software inspections are an effective way of doing this.


G. O'Regan, *Concise Guide to Software Testing*,
Undergraduate Topics in Computer Science,
https://doi.org/10.1007/978-3-030-28494-7_4

There are several approaches to inspections, and these vary in the formality of the process. An informal review consists of a walk-through of the document or code by an individual other than the author. The meeting usually takes place at the author's desk (or in a meeting room), and the reviewer and author discuss the document or code informally.

There are formal software inspection methodologies such as the well-known *Fagan inspection* methodology (Fagan 1976) and the Gilb methodology (Gilb and Graham 1994). These methodologies include pre-inspection activity, an inspection meeting, and post-inspection activity. Several inspection roles are typically employed, including an *author* role, an *inspector* role, a *tester* role, and a *moderator* role.

The Fagan inspection methodology was developed by Michael Fagan (Fig. 4.1) at IBM in the mid-1970s, and Gilb's approach to software inspections was developed by Tom Gilb in the early 1990s. The formality of the software inspection methodology employed is influenced by the impacts of software failure on the customer's business. For example, an incorrect one-line change to telecommunication software could lead to a major telecommunication outage and significant disruption to customers.

Further, there may be financial impacts resulting from failure, as a service-level agreement provides details of the service level that will be provided, and the compensation for service disruption. Consequently, a telecommunication company needs to ensure that its software is fit for purpose, and a formal software inspection process is often employed to ensure that quality is built into the software. This means that requirement documents, design documents, software code, and test documents are all inspected, and these activities need to be included in the project schedule.

The organization needs to define an inspection process that is appropriate to its business, and it may adopt a rigorous approach such as the Fagan or Gilb methodology, or a less formal review process where the impact of a failure is less severe. It may not be possible to have all of the participants physically in a room, and it may be necessary to include some reviewers via conference call or a video link. It may be more appropriate to employ a structured walk-through or informal reviews for some organizations.

**Fig. 4.1** Michael Fagan

Software inspections play an important role in building quality into the software, and the quality of the delivered software product is only as good as the quality at the end each phase, and so a phase should be exited only when the desired quality has been achieved.

The effectiveness of an inspection is influenced by the expertise of the inspectors, adequate preparation by the inspectors, the speed in which the inspection is performed, and compliance with the inspection process. A formal inspection methodology provides guidelines on the inspection and preparation rates, and entry and exit criteria are defined for the inspection.

There are generally at least two roles involved in the inspection. These are the *author* role and the *inspector* role. The *moderator, tester,* and the *reader* roles may also be present in the methodology.

Next, we describe the benefits of software inspections, and we then discuss a simple review methodology, a structured walk-through, a semi-formal review process, and the Fagan inspection process.

## 4.2 Economic Benefits of Software Inspections

There is clear evidence that a software inspection program provides a return on investment and has tangible benefits in terms of quality, productivity, time to market, and customer satisfaction. For example, IBM Houston employed software inspections for the Space Shuttle missions: 85% of the defects were found by inspections, and 15% were found by testing. There were no defects found on the space missions, and about 2 million lines of computer software were inspected. IBM, North Harbour in the UK, quoted a 9% increase in productivity with 93% of defects found by software inspections.

*Software inspections are useful for educating new employees* on the product and on the standards and procedures used in the organization. They ensure that knowledge is shared among the employees, rather than understood by just one individual. Inspections improve software productivity, as less time is spent in correcting defective software.

The cost of correction of a defect increases the later that it is identified in the lifecycle. Boehm (1981) states that the *cost of correction of a requirement defect identified in the field is over 40 times more expensive than if it were detected at the requirement phase*. Therefore, it is most economical to detect and fix defects in phase rather than correcting them later in the development cycle. The cost of correction of a requirement defect identified at the customer site includes the cost of correcting the requirements, the cost of design, coding, unit testing, system testing, and regression testing. It may be necessary to send an engineer on site to fix the problem, and there may be hidden costs in the negative perception of the company with a subsequent loss of sales.

Therefore, it is important to identify defects as early as possible, and software inspections are a cost-effective way of doing this. The *cost of poor quality* (COPQ) in an organization (Fig. 1.9) may be determined, and it involves computing the cost of internal and external failure, and the cost of appraisal and prevention.

The return on investment from the introduction of software inspections may be calculated, and the evidence is that it leads to reductions in the cost of poor quality. That is, inspections provide a cost-effective way of improving quality and productivity.

## 4.3 Informal Reviews

This type of review involves reviewers sending comments directly to the author (e.g. email or written), and there is no actual review meeting. It helps in identifying some of the defects in the work products, but its success is dependent on the extent to which the reviewers are proactive in sending comprehensive comments to the author by the due date.

The author is responsible for making sure that the review happens, and advises the participants that comments are due by a certain date. The author analyses the comments received, makes the required changes, and circulates the document for approval. The activities are described in Table 4.1.

COMMENT:
*The informal review process is dependent on the participants adequately reviewing the deliverable and sending comments to the author. The author can only request the reviewer to send comments. There is no independent monitoring of the author to ensure that the review actually happens and is effective, and that comments are requested, received, and implemented.*

**Table 4.1** Informal review

| Step | Description |
|---|---|
| 1. | The author circulates the deliverable (either physically or electronically) to the review audience |
| 2. | The author advises the review audience of the due date for comments |
| 3. | The due date for comments is typically one week or longer, and the author may send a reminder to the reviewers |
| 4. | The author checks that all comments have been received by the due date |
| 5. | The author contacts any reviewers who have not provided feedback, and requests comments |
| 6. | The author analyses all comments received and implements the appropriate changes |
| 7. | The deliverable is circulated to the review audience for sign-off |
| 8. | The reviewers sign off (with any final comments) indicating that the document has been correctly amended by the author |
| 9. | The author keeps a record of the comments received |

## 4.4 Structured Walk-through

A structured walk-through is a peer review in which the author of a deliverable (e.g. a project document or actual code) brings one or more reviewers through the deliverable. The objective is to get feedback from the reviewers on the quality of the document or code, and to familiarize the review audience with the author's work. The walk-through includes several roles, namely the *review leader* (usually the author), the *author*, the *scribe* (usually the author), and the *review audience*. It is described in more detail in Table 4.2.

## 4.5 Semi-formal Review Meeting

A semi-formal review (a simplified version of the Fagan inspection) is a moderated review meeting chaired by the review leader. The author selects the reviewers and appoints a review leader (who may be the author). The review leader chairs the meeting and verifies that the follow-up activity has been completed. The author distributes the deliverable to be reviewed and provides a brief overview of the material. This section is adapted from O'Hara (1998).

The review leader schedules the review meeting with the reviewers (with possible participation via a conference call). The review leader chairs the meeting and is responsible for keeping the meeting focused and running smoothly, resolving any conflicts, recording actions, and completing the review form.

The review leader checks that all participants (including conference call participants) are present, and that they have done sufficient preparation. Each reviewer is invited to give general comments, which will determine whether the deliverable is ready to be reviewed and whether the review should take place. Participants who

**Table 4.2** Structured walk-throughs

| Step | Description |
|---|---|
| 1. | The author circulates the deliverable (either physically or electronically) to the review audience |
| 2. | The author schedules a meeting with the reviewers |
| 3. | The reviewers familiarize themselves with the deliverable |
| 4. | The review leader (usually the author) chairs the meeting |
| 5. | The author brings the review audience through the deliverable, explaining what each section is aiming to achieve and requesting comments from them as to its correctness |
| 6. | The scribe (usually the author) records errors, decisions, and any action items |
| 7. | A meeting outcome is agreed, and the author addresses all agreed items. If the meeting outcome is that a second review should be held, then go to step 1 |
| 8. | The deliverable is circulated to reviewers for sign-off, and the reviewers sign off (with any final comments) indicating that the author has correctly amended the deliverable |
| 9. | The author keeps a record of the comments and sign-offs |

are unable to attend are required to send their comments to the review leader prior to the review, and their comments are presented at the meeting.

The material is typically reviewed page per page for a document review, and each reviewer is invited to comment on the current page. Code reviews may focus on coding standards or on both coding standards and finding defects in the software code. The issues noted during the review are recorded, and these may include items requiring further investigation.

The review outcome is decided at the end of the review (i.e. whether the deliverable needs a second review). The author then carries out the necessary corrections and investigation, and the review leader verifies that the follow-up activities have been completed. The document is then circulated to the review audience for sign-off.

COMMENT:
*The semi-formal review process works well when the review leader is not the author, as otherwise there is a risk that the review may be ineffective and that the*

**Table 4.3** Activities for semi-formal review meeting

| Phase | Review task | Roles |
|---|---|---|
| Planning | Ensure document/code is ready to be reviewed<br>Appoint *review leader* (may be author)<br>Select reviewers with appropriate knowledge/experience and assign roles | Author<br>Leader |
| Distribution | Distribute document/code and other materials to reviewers (at least 3 days before the meeting)<br>Schedule the meeting | Author<br>Leader |
| Optional meeting | Give overview of deliverable to be reviewed<br>Allow reviewers to ask any questions | Author<br>Reviewers |
| Preparation | Read through document/code, marking up issues/questions<br>Mark minor issues on their copy of the document/code | Reviewers |
| Review meeting | Review leaders chair the meeting<br>Explain purpose of the review and how it will proceed<br>Set time limit for meeting<br>Keep review meeting focused and moving<br>Review document page by page<br>Code reviews may focus on coding standards and/or identifying defects<br>Resolve any conflicts or defer to investigate<br>Note comments/shortcomings on review form<br>**Raise issues—*(do not fix them*)**<br>Reviewers make comments/suggestions/questions<br>Reviewers pass review documents/code with marked-up minor issues directly to the author<br>Author responds to any questions or issues raised<br>Propose outcome of review meeting<br>Review leader completes review summary form<br>Keep a record of the review form | Leader<br>Reviewers<br>Author |
| Post-review | Investigate and resolve any issues/shortcomings identified at review<br>Verify that the author has made the required corrections | Author<br>Leader |

**Date** ______ **Deliverable** ___________________ **Version No.** ______ **#Reviews** _____
**Author** ______________________ **Review Leader** _________________________________
**Reviewers**_______________________________________________________________

| Page/Line No. | Description | Action |
|---|---|---|
| | | |

**Unresolved Issued / Investigates**

| Issue | Reason unresolved | Verified. |
|---|---|---|
| | | |

**Review Outcome (Tick)**
No changes required □ Verification by Review Leader only □ Full review required □
Review incomplete □

**Review Summary (Optional)**
#Major Defects_______ # Minor Defects ______ Estimated Rework time ______
# Hours Preparation _______ #Hours Review _____ Amount Reviewed _______

**Fig. 4.2** Template for semi-formal review

*follow-up activity may not be done. It may work with the author acting as review leader provided the author has received the right training on the review process and consistently follows the process.*

The process for semi-formal reviews is summarized in Table 4.3. Figure 4.2 presents a template to record the issues identified during the review.

## 4.6 Fagan Inspections

The Fagan methodology (Fig. 4.3) is a well-known software inspection methodology that was developed at IBM in the mid-1970s. It is a seven-step process that includes planning, overview, preparation, an inspection meeting, process improvement, rework, and follow-up activities. Its objectives are to identify and remove errors in the work products, and to identify any systemic defects in the processes used to create the work products.

The Fagan inspection process stipulates that requirement documents, design documents, source code, and test plans are all formally inspected by experts independent of the author. The inspection is conducted from different viewpoints such as requirements, design, and test. (Table 4.4).

There are several roles defined in the inspection process, including the *moderator*, who chairs the inspection; the *reader*, who paraphrases the particular deliverable; the *author*, who is the creator of the deliverable; and the *tester*, who is concerned with the testing viewpoint. The process will consider whether the design is correct with respect to the requirements and whether the source code is correct with respect to the design.

**Fig. 4.3** Example of an inspection meeting (public domain)

**Table 4.4** Overview Fagan inspection process

| Activity | Role(s) | Objective |
|---|---|---|
| Planning | Moderator | Identify inspectors and roles<br>Verify material is ready for inspection<br>Distribute inspection material<br>Book a room for the inspection |
| Overview (optional) | Author<br>Inspectors | Brief participants on material<br>Give background information to inspectors |
| Preparation | Inspectors | Prepare for the meeting and role<br>Checklist may be employed<br>Read through the deliverable and markup issues/questions |
| Inspection meeting | Moderator/inspectors | The moderator will cancel the inspection if inadequate preparation is done<br>Time limit set for inspection<br>Moderator keeps meeting focused<br>The inspectors perform their roles<br>Emphasis is on finding defects not solutions<br>Defects are recorded and classified<br>Author responds to any questions<br>The duration of the meeting is recorded<br>An inspection outcome is agreed |
| Process improvement | Inspectors | Continuous improvement of development and inspection process<br>The causes of major defects are recorded<br>Root cause analysis to identify any systemic defect with development/inspection process<br>Recommendations are made to the process improvement team |
| Rework | Author | The author corrects the defects and carries out any necessary investigations |
| Follow-up | Moderator/author | The moderator verifies that the author has resolved the defects and investigations |

The goal is to identify as many defects as possible and to confirm the correctness of the particular deliverable. Inspection data is recorded and may be used to determine the effectiveness of the project (or organization) in detecting and preventing defects.

The moderator records the defects identified during the inspection and classifies them according to their type and severity. The defect data may be entered into an inspection database to enable analysis to be performed and metrics to be generated. The severity of the defect is recorded, and the major defects are classified [e.g. according to the Fagan defect classification or some other scheme such as the *orthogonal defect classification* (ODC)].

The next section describes the Fagan inspection guidelines, which include recommendations on the time to spend on the various inspection activities. These guidelines are very strict, and an organization may need to tailor the Fagan inspection process to suit its needs. Any tailoring of the process and guidelines need empirical evidence to confirm that they are effective.

### 4.6.1 Fagan Inspection Guidelines

The Fagan inspection guidelines provide recommendations on the amount of time that should be devoted to the various inspection activities. It is important to spend sufficient time on preparation, and that the inspection meeting is not rushed and does not attempt to cover an excessive amount of material. We first present the strict Fagan guidelines as defined by the Fagan methodology (Table 4.5) and then consider more relaxed guidelines that have been shown to be effective in the telecommunication domain (Table 4.6).

The effort involved in adherence to the strict Fagan guidelines is substantial and led to the development of tailored guidelines. The tailoring of any methodology requires care, and empirical evidence is needed to demonstrate the effectiveness of

**Table 4.5** Strict Fagan inspection guidelines

| Activity | Area | Amount/Hr | Max/Hr |
|---|---|---|---|
| Preparation time | Requirements | 4 pages | 6 pages |
| | Design | 4 pages | 6 pages |
| | Code | 100 LOC | 125 LOC |
| | Test plans | 4 pages | 6 pages |
| Inspection time | Requirements | 4 pages | 6 pages |
| | Design | 4 pages | 6 pages |
| | Code | 100 LOC | 125 LOC |
| | Test plans | 4 pages | 6 pages |

**Table 4.6** Tailored (relaxed) Fagan inspection guidelines

| Activity | Area | Amount/Hr | Max/Hr |
|---|---|---|---|
| Preparation time | Requirements | 10–15 pages | 30 pages |
| | Design | 10–15 pages | 30 pages |
| | Code | 300 LOC | 500 LOC |
| | Test plans | 10–15 pages | 30 pages |
| Inspection time | Requirements | 10–15 pages | 30 pages |
| | Design | 10–15 pages | 30 pages |
| | Code | 300 LOC | 500 LOC |
| | Test plans | 10–15 pages | 30 pages |

the tailored process (e.g. a pilot prior to its deployment which includes quantitative data to show that the inspection is effective with a low number of escaped defects).

It is important to comply with the guidelines once they are defined, and trained moderators and inspectors will ensure awareness and adherence to the methodology. Audits may be employed to verify compliance.

The tailored guidelines are presented in Table 4.6.

### 4.6.2 Inspectors and Roles

There are four inspector roles identified in a Fagan inspection (Table 4.7).

### 4.6.3 Inspection Entry Criteria

Entry criteria for the various types of inspections are specified in Table 4.8 and should be satisfied for an effective inspection.

**Table 4.7** Inspector roles

| Role | Responsibilities |
|---|---|
| Moderator | Manages the inspection process and ensures compliance with the process<br>Trained in the inspection process and as a moderator<br>Skilful, diplomatic, and occasionally forceful<br>Plans the inspection and chairs the meeting<br>Keeps to the inspection guidelines<br>Verifies that the deliverables are ready to be inspected<br>Verifies that the inspectors have done adequate preparation<br>Keeps the meeting focused and resolves any conflicts<br>Records the defects on the inspection sheet<br>Verifies that the agreed follow-up work has been completed |
| Reader | Paraphrases the deliverable and gives an independent view of it<br>Actively participates in the inspection |
| Author | Creator of the work product being inspected<br>Has an interest in finding all defects present in the deliverable<br>Ensures that the work product is ready to be inspected<br>Gives an overview to inspectors (if required)<br>Participates actively during inspection and answers all questions<br>Resolves all identified defects and carries out required investigation |
| Tester | Role is focused on how the product would be tested<br>Role often employed in requirement inspection/test plan inspection<br>The tester participates actively in the inspection |

**Table 4.8** Fagan entry criteria

| Inspection type | Entry criteria | Roles |
|---|---|---|
| Requirements | Inspector(s) with sufficient expertise available<br>Correct requirement template used<br>Preparation done by inspectors | Moderator/inspectors |
| Design inspection | Requirements inspected and signed off<br>Inspector(s) have sufficient domain knowledge<br>Correct design template used<br>Preparation done by inspectors | Moderator/inspectors |
| Code inspection | Requirements/design inspected and signed off<br>Overview provided<br>Code listing available<br>Clean compilation of source code<br>Coding standards satisfied<br>Inspector(s) have sufficient domain knowledge<br>Preparation done by inspectors | Moderator/ inspectors |
| Test plan inspection | Requirements/design inspected and signed off<br>Inspector(s) have sufficient domain knowledge<br>Correct test plan template employed<br>Preparation done by inspectors | Moderator/ inspectors |

### 4.6.4 Preparation

Preparation is a key part of the process, as the inspection will be ineffective if insufficient time is devoted to understanding and reviewing the deliverables prior to the inspection meeting. Consequently, the moderator is required to cancel the inspection if the inspectors are insufficiently prepared.

### 4.6.5 The Inspection Meeting

The inspection meeting (Table 4.9) consists of a formal meeting between the author and at least one inspector. It is concerned with finding defects in the particular deliverable and verifying its correctness. The effectiveness of the inspection is influenced by

– The expertise and experience of the inspector(s)
– Preparation done by inspector(s)
– The speed of the inspection.

These factors are quite clear since an inexperienced inspector will lack the appropriate domain knowledge to understand the material in depth. Second, an inspector who is inadequately prepared will be unable to make a substantial contribution during the inspection. Third, the inspection is ineffective if it tries to cover

**Table 4.9** Inspection meeting

| Inspection type | Purpose | Procedure |
|---|---|---|
| Requirements | Find requirement defects<br>Confirm requirements correct | Inspectors review each page of requirements and raise questions or concerns. Defects recorded by moderator |
| Design | Find defects in design<br>Confirm correct (with respect to requirements) | Inspectors review each page of design (compare to requirements) and raise questions or concerns. Defects recorded by moderator |
| Code | Find defects in the code<br>Confirm code correct (with respect to design/requirements) | Inspectors review the code, compare to requirements/design, and raise questions or concerns. Defects recorded by moderator |
| Test | Find defects in test cases/test plan<br>Confirm test cases sufficient to verify design/requirements | Inspectors review each page of test plan/test specification, compare to requirements/design, and raise questions or concerns. Defects recorded by moderator |

too much material in a short space of time. The moderator will complete the inspection form to record the results from the inspection (Fig. 4.5).

The next part of the inspection is concerned with process improvement. The inspector(s) and author examine the major defects, identify the root causes of the defect, and determine corrective action to address any systemic defects in the software process.

The outcome of the inspection is agreed (e.g. inspect the material again or verification by the moderator). The moderator is responsible for completing the inspection summary form and the defect log form, and for entering the inspection data into the inspection database. The moderator will give any process improvement suggestions directly to the process improvement team. The author then makes the agreed changes, and these are verified by the moderator (or a re-inspection).

### 4.6.6 Inspection Exit Criteria

The exit criteria (Table 4.10) for the various inspections are as follows.

### 4.6.7 Issue Severity

The severity of an issue identified in the Fagan inspection may be classified as major, minor, a process improvement item, or an item requiring further investigation (Table 4.11). It is classified as *major* if its non-detection would lead to a defect report being raised later in the development cycle, whereas a defect report would

**Table 4.10** Fagan exit criteria

| Inspection type | Exit criteria |
|---|---|
| Requirements | Requirements satisfy the customer's needs<br>All requirement defects are corrected |
| Design | Design satisfies the requirements<br>Design satisfies the design standards<br>All identified defects are corrected |
| Code | Code satisfies the design and requirements<br>Code satisfies coding standards and compiles cleanly<br>All identified defects corrected |
| Test | Test plan/specification sufficient to test the requirements/design<br>Test plan/specification follows test standards<br>All identified defects corrected |

**Table 4.11** Issue severity

| Issue severity | Definition |
|---|---|
| Major (M) | A defect in the work product that would lead to a customer-reported problem if undetected |
| Minor (m) | A minor issue in the work product |
| Process improvement (PI) | A process improvement suggestion based on analysis of major defects |
| Investigate (INV) | An item to be investigated |

generally not be raised for a *minor* issue. An issue classified as an investigate item requires further study, and an issue classified as process improvement is used to improve the software development process.

### 4.6.8 Defect Type

There are several defect-type classification schemes employed in software inspections. These include the Fagan inspection defect classification (Table 4.12) and the orthogonal defect classification scheme (Table 4.13).

The orthogonal defect classification (ODC) scheme was developed at IBM (Bhandari 1993), and it classifies a defect into three orthogonal viewpoints. The *defect trigger* is the catalyst that led the defect to manifest itself; the *defect type* indicates the change required for correction; the *defect impact* indicates the impact of the defect at the phase in which it was identified. The ODC yields a rich pool of information about the defect, but effort is required to record this information. The defect-type classification is described in Table 4.13.

The defect impact provides a mechanism to relate the impact of the software defect to customer satisfaction. The impact of a defect-identified pre-release is

**Table 4.12** Classification of defects in Fagan inspections

| Code inspection | Type | Design inspections | Type | Requirement inspections | Type |
|---|---|---|---|---|---|
| Logic (code) | LO | Usability | UY | Product Objectives | PO |
| Design | DE | Requirements | RQ | Documentation | DS |
| Requirements | RQ | Logic | LO | Hardware interface | HI |
| Maintainable | MN | Systems Interface | IS | Completion analysis | CO |
| Data usage | DA | Portability | PY | Function | FU |
| Performance | PE | Reliability | RY | Software interface | SI |
| Standards | ST | Maintainability | MN | Performance | PE |
| Code Comment | CC | Error handling | EH | Reliability | RL |
| | | Other | OT | Spelling | GS |

**Table 4.13** Classification of ODC defect types

| Defect type | Code | Definition |
|---|---|---|
| Checking | CHK | Omission or incorrect validation of parameters or data in conditional statements |
| Assignment | ASN | Value incorrectly assigned or not assigned at all |
| Algorithm | ALG | Efficiency or correctness issue in algorithm |
| Timing | TIM | Timing/serialization error between modules and shared resources |
| Interface | INT | Interface error (error in communications between modules, operating system, etc.) |
| Function | FUN | Omission of significant functionality |
| Documentation | DOC | Error in user guides, installation guides, or code comments |
| Build/merge | BLD | Error in build process/library system or version control |
| Miscellaneous | MIS | None of the above |

viewed as the impact of it being detected by an end user, and for a customer-reported defect its impact is the actual information reported by the customer.

The inspection data is generally recorded in an inspection database, which allows analysis to be performed on the most common types of defects. The frequency of defects per category is identified, and causal analysis is employed to identify preventive actions (Fig. 4.4). The most problematic areas are targeted first (as identified in a Pareto chart), and an investigation into the causes of a particular category is conducted. Action plans will be prepared to improve the existing processes to prevent reoccurrence.

The ODC scheme may be used to give early warning on the quality and reliability of the software, as its use leads to an expected profile of defects for the various lifecycle phases. The actual profile may then be compared to the expected profile, and significant differences may indicate risks to quality.

For example, if the actual defect profile at the system test phase resembles the defect profile of the unit testing phase, then this may indicate quality problems. The

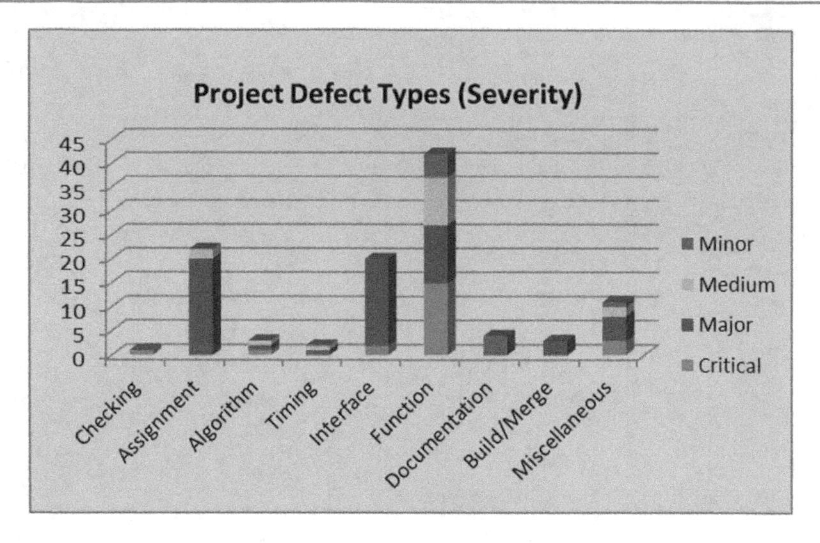


**Fig. 4.4** Defect types in a project (ODC)

unit testing phase is expected to yield a certain pool of defects, with system testing receiving higher-quality software with the defects found during unit testing removed. Consequently, ODC may be applied to make a judgment of product quality and performance.

The inspection data will enable the *phase containment effectiveness* (PCE) metric to be determined (Fig. 9.10) and to determine if the software is ready for release to the customer.

## 4.7 Automated Code Inspections

Static code analysis is the analysis of software code without the execution of the code, and it is often performed with automated tools. The actual analysis done depends on the sophistication of the tool, with some tools analysing individual statements or declarations, whereas others may analyse the whole source code. The objective of the analysis is to highlight potential coding errors early in the software development lifecycle (Fig. 4.5).

Compilers find defects in the syntax of the software code, whereas static code analysis tools can analyse the software code to find more defects. Static code analysis provides early detection of defects prior to test execution, as well as making the software code easier to maintain. They also help in identifying defects that are hard to find in dynamic testing, and they may check for:

– Violation of coding standards
– Never-ending loops

**Inspection Type** ________ **Deliverable** ______________ **Project** ____________
**Date** ________________ **Amount Inspected** _____ **Version No.** ____
**Author**________________ **Moderator**______________ **No. of Reviews** _____
**Inspectors** ________________________________________
**#Hours Preparation** ________ **# Hours Inspection** _________ **#Hours Rework** _____
**Summary of Findings:** **# Majors** ____ **# Minors** ___ **# PIs** _____ **# INVs** ___
**ODC Summary (Majors): #CHK__ #ASS__ #ALG__ #TIM__ # INT__ #FUN___ # DOC__ # BLD__**

**No. Page/Line No. Severity Type Description**

**Top 3 Root Causes of Major Defects / Process Improvement Actions**
1.
2.
3.

**Review Outcome**
No changes □ Verification by Moderator □ Full Review □ Review Incomplete □
Defects per KLOC _____ Defects per page _____ Verification of Rework ______
Date Verified ________ Inspection Data in Database ____

**Fig. 4.5** Template for Fagan inspection

– Unreachable code
– Variables that are never used
– Referencing a variable with an undefined value
– Security vulnerabilities
– Parameter-type mismatch.

The automated software inspection tools provide quality assessment reports on the extent to which the coding standards are satisfied, as well as quality metrics on code complexity (Fig. 10.2). They provide warnings about potentially problems with the complexity of the design with metrics that have a high complexity measure. Many integrated development environments (IDEs) provide basic functionality for automated code reviews.

They provide metrics on the maintainability of the code, and some tools give a visual picture of system complexity and include a re-factoring feature to assist in reducing complexity. They automatically generate code assessment reports listing all of the files examined and provide metrics on the clarity, maintainability, and testability of the code.

The compliance with coding standards is important in producing readable and maintainable code, and in preventing error-prone coding styles. There are several tools available to check conformance to coding standards, which include reporting capabilities to show code quality as well as fault detection and avoidance measures.

## 4.8 Review Questions

1. What are software inspections?
2. Explain the difference between informal reviews, structured walk-throughs, semi-formal reviews, and formal inspections.
3. Explain the difference between static testing and dynamic testing
4. Describe the seven steps in the Fagan inspection process.
5. What is the purpose of entry and exit criteria in software inspections?
6. What factors influence the effectiveness of a software inspection?
7. Describe the roles involved in a Fagan inspection.
8. Describe the benefits of automated inspections.
9. What are the benefits of software inspections?

## 4.9 Summary

The objective of software inspections is to build quality into the software product, as the cost of correction of a defect increases later in the software development cycle in which it is detected. They make economic sense as it is more cost effective to build quality in rather than adding it later in the development cycle.

There are several approaches to software reviews and inspections, including a walk-through of the document or code by an individual than the author. This meeting is informal and usually takes place at the author's desk or in a meeting room, and the reviewer and author discuss the document or code informally.

There are formal software inspection methodologies such as the well-known Fagan inspection methodology. This approach includes pre-inspection activity, an inspection meeting, and post-inspection activity. Several inspection roles are typically employed, including an author role, an inspector role, a tester role, and a moderator role.

The level of formality of an inspection process is influenced by the business and the potential impact of a software defect on its customers. It may not be possible to have all of the participants physically present in a room, and participation by conference call may be employed.

The effectiveness of an inspection is influenced by the expertise of the inspectors, adequate preparation, and speed of the inspection, and compliance with the inspection process.

Static code analysis is the analysis of software code without executing the code, and it is usually performed with specialized automated testing tools. The objective is to highlight potential coding errors early in the software development lifecycle.

## References

Bhandari I (1993) A case study of software process improvement during development. IEEE Trans Softw Eng 19(12)
Boehm B (1981) Software engineering economics. Prentice Hall, New Jersey
Fagan M (1976) Design and code inspections to reduce errors in software development. IBM Syst J 15(3)
Gilb T, Graham D (1994) Software inspections. Addison Wesley, Boston
O'Hara F (1998) Peer reviews—the key to cost effective quality. European SEPG, Amsterdam

# Software Test Planning 5

**Key Topics**

Estimation
Work breakdown structure
Scheduling
Risk management
Project governance
Test reporting
Test monitoring and control

## 5.1 Introduction

Testing is a sub-project of a project and needs to be managed as such, and so good planning and monitoring and control are required. Test planning involves defining the scope of the testing to be performed; defining the test environment; estimating the effort required to define the test cases and to perform the testing; identifying the resources needed (including people, hardware, software, and tools); assigning the resources to the tasks; defining the schedule; and identifying any risks to the testing and managing them.

Test monitoring and control involve monitoring progress and taking corrective action when progress deviates from expectations; re-planning where the scope of the testing has changed; communicating progress to the various stakeholders with test reports to provide visibility into the testing carried out; taking corrective action to ensure quality and schedule are achieved; managing risks and issues; managing the change requests that arise during the project; and providing a final test report



G. O'Regan, *Concise Guide to Software Testing*,
Undergraduate Topics in Computer Science,
https://doi.org/10.1007/978-3-030-28494-7_5

with a recommendation to go to acceptance testing. The effective management of testing involves:

- Defining the scope of the testing
- Determining types of testing to be performed
- Estimates of time, effort, cost, resources, people, hardware, software, and tools
- Determining the start and end dates for the testing
- Determining the resources and staffing required
- Determining how test progress will be communicated
- Defining how test defects will be logged and reported
- Definition of test environment
- Assigning resources to the various tasks and activities
- Preparing the test plan
- Scheduling the various tasks and activities
- Preparing the initial test schedule and key milestones
- Identifying the key risks to testing
- Monitoring progress, budget, schedule, effort, risks, issues, change requests, and quality and taking corrective action
- Re-planning and rescheduling
- Providing regular status of passed, blocked, failed tests
- Communicating progress to affected stakeholders
- Preparing status reports and presentations
- Re-planning if scope of the project changes
- Conducting post-mortem to learn any lessons from the testing.

The test plan for the project is documented (this could be part of the project plan, but it is often in a separate document based on the test planning template in the IEEE 829 standard). It includes the scope of the testing, the personnel involved, the resources and effort required, the key milestones, the definition of the test environment, any special hardware and test tools required, and the planned test schedule. There is a separate test specification plan for the various types of testing, which records the test cases, including the purpose of each test case, the inputs and expected outputs, and the test procedure for the execution of the particular test case.

Several types of testing are performed during the project, including unit, integration, system, regression, performance, and user acceptance testing. The software developers perform the unit testing to verify the correctness of a module. This type of testing is termed "*white box*" testing and is based on knowledge of the internals of the software module. It involves defining and executing test cases to ensure code and branch coverage. The objective of "*black box*" testing is to verify the functionality of a module (or feature or the complete system itself), and knowledge of the internals of the software module is not required.

Test reporting is an important part of the project, and it ensures that all project participants understand the current quality of the software, as well as understanding what needs to be done to ensure that the product achieves the desired quality criteria. The test status is reported regularly during the project, and once the tester discovers a defect, a problem report is opened, and the problem is analysed and corrected by the software developers. The problem may indicate a genuine defect, a misunderstanding by the tester, or a request for an enhancement.

Table 3.3 presents a simple test schedule for a small project, and Microsoft Project is generally employed for the planning and tracking of larger projects (Fig. 5.2). The activities in the test schedule are tracked and progress updated to record the tasks that have been completed, with new dates applied to tasks that have fallen behind schedule. Testing is a key sub-project of the main project, and the project manager will track the key test milestones and will maintain close contact with the test manager.

The effective management of risk during testing is essential to project success. It is prudent to consider risk management early in test planning, to identify risks that could potentially arise during the testing, and to identify (as far as is practical) actions to mitigate the risk or a contingency plan to address the risk if it materializes.

Risks arise due to uncertainty, and the risk management cycle involves[1] risk identification; risk analysis and evaluation; identifying responses to risks; selecting and planning a response to the risk; and risk monitoring. Once the risks have been identified, they are logged (e.g. in the risk log). The likelihood of each risk arising and its impact should it materialize is then determined. The risk is assigned an owner and an appropriate response to the risk determined.

Estimation is difficult as software projects are often breaking new ground and differ from previous projects. That is, historical estimates may often not be a good basis for estimation for the current project. Often, unanticipated problems may arise for technically advanced projects, and the estimates may be overly optimistic.

*Gantt charts* are generally employed for project scheduling, and these show the work breakdown for the project as well as task dependencies and allocation of staff to the various tasks.

Two popular project management methodologies are the *PRINCE*2 methodology (Office of Government Commerce 2004), which was developed in the UK, and *Project Management Professional (*PMP*)* and its associated project management body of knowledge (PMBOK) from the *Project Management Institute* (PMI) in the USA.

The test manager works closely with the project manager during the project, with the project manager responsible for the day-to-day management of the project and the test manager responsible for the day-to-day management of the testing. The *project board* (or steering group) includes the key stakeholders and is accountable for the success of the project. The project manager provides regular status reports to the project board during the project, and the test manager liaises with the project manager to ensure that the key test status is presented. The project board is consulted when key project decisions need to be made.

[1]These are the risk management activities in the PRINCE2 methodology.

## 5.2 Test Estimation

Estimation is a key part of project planning, and the accurate estimates of effort, cost, and schedule are essential to delivering a project on time and on budget, and with the right quality.[2] Estimation is employed in the planning process to determine the resources and effort required, and it feeds into the scheduling of the testing. The problems with over- or underestimation of projects are well known, and good estimates allow:

- Accurate calculation of the cost of testing
- Accurate scheduling of the testing
- Measurement of progress and costs against the estimates
- Determining the resources required for the testing.

Poor estimation leads to:

- Testing being over- or underestimated
- Testing being over- or under-resourced (impacting staff morale)
- Negative impression of the test manager.

Consequently, estimation needs to be rigorous, and there are several well-known techniques available (e.g. work breakdown structures, function points, and so on). Estimation applies to both the early and the later parts of the project, with the later phases of the project refining the initial estimates, as a more detailed understanding of the testing is then available. The new estimates are used to reschedule and to predict the eventual effort, delivery date, and cost of the project. The following are guidelines for estimation:

- Sufficient time needs to be allowed to do estimation.
- Historical data is often employed.
- Brainstorming is often employed.
- The initial estimates are high level.
- The estimates should be conservative rather than optimistic.
- Estimates will usually include contingency.
- Estimates should be reviewed to ensure their adequacy.
- Estimates from independent experts may be useful.
- It may be useful to prepare estimates using several methods and to compare.

Project metrics for testing (Figs. 9.4 and 9.5) may be employed to measure the accuracy of the estimates for the test planning. These include:

[2]The consequences of underestimating a project include the project being delivered late, with the project team working late nights and weekends to recover the schedule, quality being compromised with steps in the process omitted, and so on.

– Effort estimation accuracy
– Budget estimation accuracy
– Schedule estimation accuracy.

Next, we discuss several estimation techniques including the work breakdown structure, the analogy method, and the Delphi method.

### 5.2.1 Estimation Techniques

Estimates need to be produced consistently, and it would be inappropriate to have an estimation procedure such as *"Go ask Fred"*[3], as this clearly relies on an individual and is not a repeatable process. The estimates may be based on a work breakdown structure, function points, or another appropriate methodology. There are several approaches to estimation (Table 5.1) including.

### 5.2.2 Work Breakdown Structure

This is a popular approach to estimation (*it is also known as decomposition*) and involves the following:

– Identify the deliverables to be produced during the testing.
– Estimate the size of each deliverable (in pages or #test cases).
– Estimate the effort (number of days) required to complete the deliverable based on its complexity and size, and experience of team.
– Estimate the cost of the completed deliverable.
– The estimate for the testing is the sum of the individual estimates.

The approach often uses productivity data that is available from previously completed projects. The effort required for a complex deliverable is higher than that of a simple deliverable (where both are of the same size). The test planning section of the project plan (or a separate test plan) will detail the deliverables/tasks to be carried out in each phase. It may include a table similar to Table 5.2.

## 5.3 Test Planning and Scheduling

A well-managed project has an increased chance of success, and good planning is an essential part of project management. There is the well-known adage that states *"Fail to plan, plan to fail"*[4]. The test manager and the relevant stakeholders will

[3]Unless "Go ask Fred" is the name of the estimation methodology or the estimation tool employed.
[4]This quotation is adapted from Benjamin Franklin (an inventor and signatory to the American declaration of independence).

**Table 5.1** Estimation techniques

| Technique | Description |
|---|---|
| Work breakdown structure | Identify the deliverables to be produced during the testing. Estimate the size of each deliverable (in pages or #test cases). Estimate the effort (number of days) required to complete the deliverable based on its size and complexity. Estimate the cost of the completed deliverable |
| Analogy method | This involves comparing the proposed testing with a previously completed project (that is similar to the proposed project). The historical data and metrics for schedule, effort, and budget estimation accuracy are considered, as well as similarities and differences between the projects to provide effort, schedule, and budget estimates |
| Expert judgment | This involves consultation with experienced personnel to derive the estimate. The expert(s) can factor in differences between past projects, knowledge of existing systems as well as the specific requirements of the testing |
| Delphi method | The *Delphi method* is a consensus method used to produce accurate schedules and estimates. It was developed by the RAND Corporation and improved by Barry Boehm and others. It provides extra confidence in the estimates by using experts independent of the test manager |
| Cost predictor models | These include various cost prediction models such as *COCOMO* and SLIM. The Costar tool supports COCOMO, and the Qsm tool supports SLIM |
| Function points | *Function points* were developed by Allan Albrecht at IBM in the late 1970s and involve analysing each functional requirement and assigning a number of function points based on its size and complexity. This total number of function points is a measure of the estimate for the testing |

consider the appropriate approach for the testing and determine whether the testing should be outsourced to a third-party supplier or whether the test group has the competence and resources to perform the testing internally. A simple process map for test planning is presented in Fig. 5.1.

The effort estimates are used in scheduling of the tasks and activities using a project-scheduling tool such as *Microsoft Project* (Fig. 5.2). The schedule will include the key test milestones, the activities and tasks to be performed as well as their associated timescales, and the resources required to carry out each task.

The test manager will create the project test plan and schedule, and track the schedule to completion. The test manager will update the project schedule regularly during the project. The project test plan defines how the testing will be carried out, and it generally includes sections such as:

– Scope of testing
– Types of testing to be performed
– Roles and responsibilities
– Key stakeholders
– Resources required (hardware and human)
– Training, knowledge, and skills required

**Table 5.2** Example of work breakdown structure for test estimation

| Lifecycle phase | Project deliverable or task description | Est. size | Est. effort | Est. cost |
|---|---|---|---|---|
| Requirements/design | Inspections | | 2 days | $1000 |
| Coding phase | Code inspections | 50 | 2 days | $1000 |
| | Unit test cases | | 2 days | $1000 |
| | Unit testing | | 2 days | $1000 |
| Testing | Prepare test plan | 20 | 3 days | $1500 |
| | Define test environment | | 1 day | $500 |
| | Set up test environment | | 2 days | $1000 |
| | System test specs | | 2 days | $1000 |
| | Inspection | | 0.5 day | $250 |
| | System testing | | 1 day | $500 |
| | Performance test specs | | 2 days | $1000 |
| | Inspection | | 0.5 day | $250 |
| | Performance testing | | 1 day | $500 |
| | Regression tests | | 2 days | $1000 |
| | Regression testing | | 1 day | $500 |
| | UAT specs | | 2 days | $1000 |
| | Inspection | | 0.5 day | $250 |
| | UAT | | 1 day | $500 |
| | Test reporting | | 2 days | $1000 |
| Total | | | 29.5 | $14750 |
| Contingency | 10% | | | $1475 |
| Total | | | | $16,225 |

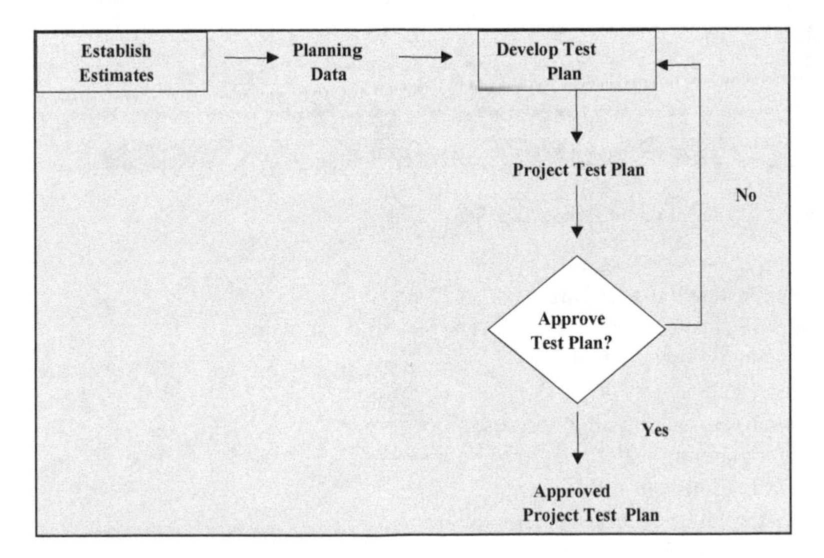


**Fig. 5.1** Simple process map for test planning

| | | WBS | Task Name | Duration | Work | Start | Finish |
|---|---|---|---|---|---|---|---|
| 1 | | 1 | **Project Management** | **277 days** | **102.55 days** | **Mon 16/01/06** | **Wed 20/12/06** |
| 2 | ✓ | 2 | **Planning** | **26.7 days** | **20 days** | **Mon 16/01/06** | **Thu 16/02/06** |
| 3 | ✓ | 2.1 | Set up Teams | 12 days | 3 days | Mon 16/01/06 | Mon 30/01/06 |
| 4 | ✓ | 2.2 | Define Team Charters | 4 days | 2 days | Tue 24/01/06 | Mon 30/01/06 |
| 5 | ✓ | 2.3 | Develop Presentations(Steering/Kick Off) | 3 days | 3 days | Mon 30/01/06 | Wed 01/02/06 |
| 6 | ✓ | 2.4 | Deliver Presentation to Steering Group | 0.25 days | 0.75 days | Wed 01/02/06 | Wed 01/02/06 |
| 7 | ✓ | 2.5 | Document Improvement Plan | 6 days | 4 days | Mon 23/01/06 | Mon 30/01/06 |
| 8 | ✓ | 2.6 | Document Schedule in MSP | 4 days | 2 days | Tue 31/01/06 | Fri 03/02/06 |
| 9 | ✓ | 2.7 | Document Software Development Policy | 7 days | 2 days | Mon 23/01/06 | Tue 31/01/06 |
| 10 | ✓ | 2.8 | Kick Off Meeting | 0.25 days | 1.75 days | Thu 16/02/06 | Thu 16/02/06 |
| 11 | ✓ | 2.9 | Define Issue Log, Risk Log and LL Log | 5 days | 1 day | Mon 23/01/06 | Fri 27/01/06 |
| 12 | ✓ | 2.10 | Define Steering Group Report Template | 2.5 days | 0.5 days | Mon 30/01/06 | Wed 01/02/06 |
| 13 | ✓ | 2.11 | Planning Complete | 0 days | 0 days | Wed 01/02/06 | Wed 01/02/06 |
| 14 | ✓ | 3 | **CMMI Training** | **6.71 days** | **11 days** | **Wed 01/03/06** | **Wed 08/03/06** |
| 15 | ✓ | 3.1 | Overview Training (Group A) | 1 day | 3 days | Wed 01/03/06 | Wed 01/03/06 |
| 16 | ✓ | 3.2 | Overview Training (Group B) | 1 day | 3 days | Thu 02/03/06 | Thu 02/03/06 |
| 17 | ✓ | 3.3 | SEPG Training | 1 day | 5 days | Wed 08/03/06 | Wed 08/03/06 |
| 18 | ✓ | 3.4 | Training Complete | 0 days | 0 days | Wed 08/03/06 | Wed 08/03/06 |
| 19 | | 4 | **CMMi Definition and Rollout** | **185.14 days?** | **215.26 days** | **Tue 07/03/06** | **Thu 19/10/06** |
| 20 | ✓ | 4.1 | **PP+PMC** | **134.14 days** | **63.08 days** | **Tue 14/03/06** | **Thu 24/08/06** |
| 21 | ✓ | 4.1.1 | Workshop to agree process | 0.5 days | 2.5 days | Tue 14/03/06 | Tue 14/03/06 |
| 22 | ✓ | 4.1.2 | Document Parts of PM Process | 6.5 days | 2.7 days | Mon 20/03/06 | Mon 27/03/06 |
| 23 | ✓ | 4.1.3 | Workshop to agree process | 0.5 days | 2.5 days | Tue 28/03/06 | Tue 28/03/06 |
| 24 | ✓ | 4.1.4 | Document Parts of PM/CMgt Process | 8 days | 3.17 days | Thu 30/03/06 | Mon 10/04/06 |
| 25 | ✓ | 4.1.5 | Workshop to agree process | 0.5 days | 2.5 days | Tue 11/04/06 | Tue 11/04/06 |
| 26 | ✓ | 4.1.6 | Updates to Processes & Templates | 10 days | 8 days | Fri 14/04/06 | Wed 26/04/06 |
| 27 | ✓ | 4.1.7 | Workshop to Agree Process | 0.5 days | 2.5 days | Tue 09/05/06 | Tue 09/05/06 |
| 28 | ✓ | 4.1.8 | Updates to Processes & Templates | 5 days | 4 days | Tue 09/05/06 | Mon 15/05/06 |
| 29 | ✓ | 4.1.9 | PM Procedure & Updates to Templates | 20 days | 20 days | Wed 31/05/06 | Fri 23/06/06 |
| 30 | ✓ | 4.1.10 | SEPG Approval | 0.5 days | 2.5 days | Tue 27/06/06 | Tue 27/06/06 |
| 31 | ✓ | 4.1.11 | Pilots | 26.79 days | 5.47 days | Wed 28/06/06 | Mon 31/07/06 |
| 32 | ✓ | 4.1.12 | Training and Rollout | 20 days | 7.25 days | Tue 01/08/06 | Thu 24/08/06 |

**Fig. 5.2** Sample Microsoft Project schedule

– Key milestones (for testing)
– Schedule (for test activities, deliverables, and estimates)
– Key assumptions
– Key risks
– Communication planning and test reporting
– Budget planning
– Defect logging and retesting
– Test acceptance criteria
– Configuration management.

**Table 5.3** Sample test planning checklist

| No. | Item to check |
|---|---|
| 1. | Is the test plan complete and approved by the stakeholders? |
| 2. | Are estimates available for testing? Are they realistic? |
| 3. | Has the change control mechanism been set up for the project? |
| 4. | Are the risk log, issue log, and lessons learned log set up? |
| 5. | Are the responses to the risks and issues appropriate? |
| 6. | Have project and test communication been appropriately planned? |
| 7. | Are the key milestones defined? |
| 8. | Is the test schedule available? |
| 9. | Is the test schedule up to date? |
| 10. | Is the testing appropriately resourced? |
| 11. | Are all deliverables under configuration management control? |

There will be dedicated test plans for unit testing, system testing, and UAT. These are generally prepared as part of test case analysis and design which is described in Chap. 6, and we describe them briefly in Sect. 5.5.

Communication planning describes how communication will be carried out during the testing, and it includes the various meetings and reports that will be produced; financial planning is concerned with budget planning for the project (including the testing); configuration management is concerned with identifying the configuration items (i.e. the test deliverables) to be controlled and systematically controlling changes to them throughout the lifecycle (see Chap. 15). It ensures that all of the project deliverables are kept consistent following approved changes during the project.

The project test plan is a key project document, and it needs to be approved by the stakeholders. The test manager needs to ensure that the project test plan, the test schedule, and technical work products are kept consistent with the requirements. In other words, if there are changes to the requirements, then the project test plan and schedule will need to be updated accordingly.

Checklists are useful in verifying that the tasks have been completed. The sample checklist below (Table 5.3) may be tailored to verify that the test planning has been appropriately performed.

## 5.4 Risk Management in Testing

Risks arises due to uncertainty, and *risk management is concerned with managing uncertainty* and especially the management of any undesired events. Risks need to be identified, analysed, and controlled in order for the project to be successful, and risk management activities take place throughout the project lifecycle. There are risks that are specific to testing, and while the project manager has overall

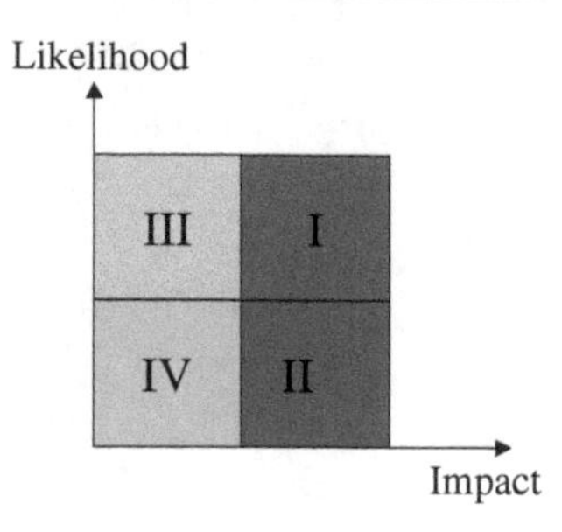


**Fig. 5.3** Risk categories

responsibility for risk management, in practice the test manager and project manager will work very closely on these risks.

Once the initial set of risks to the project has been identified, they are analysed to determine their *likelihood of occurrence* and their *impact* (e.g. on cost, schedule, or quality) should they materialize. These two parameters determine the *risk category*, and the most serious risk category refers to a risk with a high probability of occurring and a high impact on occurrence (i.e. Box I in Fig. 5.3).

Countermeasures are defined to reduce the likelihood of occurrence and impact of the risks, and contingency plans are prepared to deal with the situation of the risk actually materializing. Additional risks may arise during the testing, and the project manager/test manager need to be proactive in their identification and management.

Risks need to be reviewed regularly, especially following changes to the project. These could be changes to the business case or the business requirements, loss of key personnel, and so on. In other words, events that occur during the project may affect existing risks (including the probability of their occurrence and their impact) and may lead to new risks. Countermeasures need to be kept up to date during the project. Risks are reported regularly throughout the project.

Table 5.4 summarizes the activities in the risk management cycle including identifying risks; determining the probability of their occurrence and impact should they occur; identifying responses to the risks; and monitoring and reporting.

The project manager/test manager will maintain a risk repository (this may be a tool or a risk log) to record details of each risk, including its type and description; its likelihood and its impact (yielding the risk category); as well as the response to the risk. Sample risks to the testing in the project include:

- The software may be delivered late leading to the test schedule being cut short.
- The software may be of poor quality meaning that only limited testing may be done.
- A tester may resign.
- Specialized hardware required for testing may not arrive on time.
- The relationship between developers and testers may become antagonistic.
- The testers may lack the expertise to properly test the software.

**Table 5.4** Risk management activities

| Activity | Description |
| --- | --- |
| Risk management strategy | This defines how the risks will be identified, monitored, reviewed, and reported during the project, as well as the frequency of monitoring and reporting |
| Risk identification | This involves identifying the risks to the project and recording them in a risk repository (e.g. risk log). It continues throughout the project lifecycle. PRINCE2 classifies risks into:<br>– *Business* (e.g. collapse of subcontractors)<br>– *Legal and regulatory*<br>– *Organizational* (e.g. skilled resources/management)<br>– *Technical* (e.g. scope creep, architecture, design)<br>– *Environmental* (e.g. flooding or fires) |
| Evaluating the risks | This involves assessing the likelihood of occurrence of a particular risk and its impact (on cost, schedule, etc.) should it materialize. These two parameters result in the risk category |
| Identifying risk responses | The project manager/test manager will determine the appropriate response to a risk such as reducing the probability of its occurrence or its impact should it occur. These include:<br>– *Prevention* which aims to prevent it from occurring<br>– *Reduction* aims to reduce the probability of occurrence or impact should it occur<br>– *Transfer* aims to transfer the risk to a third party<br>– *Acceptance* is when nothing can be done about it<br>– *Contingency* is action that is carried out should the risk materialize |
| Risk monitoring and reporting | This involves monitoring existing risks to verify that the actions taken to manage the risks are effective, as well as identifying new risks. This helps in providing an early warning that an identified risk is going to materialize, and *a risk that materializes is a new project issue* that needs to be dealt with |
| Lessons learned | This is concerned with determining the effectiveness of risk management during the project and to learn any lessons for future projects |

## 5.5 Dedicated Test Plans

There will generally be specific test plans for the various types of testing performed such as unit, system, performance, and UAT. These plans specify how each type of testing will be performed, and they may include sections such as:

– Test objectives
– Approach
– Roles and responsibilities
– Key stakeholders
– Assumptions

– Risks
– Resources required
– Training required
– Preparation dates
– Testing dates
– Test environment
– Test tools
– Entry and exit criteria.

The dedicated test plan may contain a summary of the test cases to be executed as well as a traceability matrix that shows how the test cases cover the user or system requirements or design. These plans are written by the tester (or possibly the test leader depending on how the test team is organized). We discuss dedicated test plans again in Chap. 6.

## 5.6 Monitoring and Control

Test monitoring and control are concerned with monitoring test execution to give feedback and visibility on the test activities, and taking corrective action when performance deviates from expectations. The progress with the testing needs to be monitored against the plan, and corrective action taken when progress deviates from expectations. The key parameters such as effort and schedule as well as risks and issues are monitored, and the status of the testing is communicated regularly to the affected stakeholders.

The test manager will conduct regular progress and milestone reviews with the test team to determine actual progress and to identify new risks and issues. Figure 7.3 presents a simple process map for test monitoring and control, and the main focus is:

– Monitor the test plan and schedule.
– Monitor risks and issues and take appropriate action.
– Monitor resources and manage any resource issues.
– Conduct progress and milestone reviews.
– Measure test case execution, defect information, and test coverage.
– Re-plan as appropriate.
– Track corrective action to closure.
– Maintain close contact with the project manager and keep informed on progress.
– Prepare and present test reports detailing the test status.

The test manager is responsible for test monitoring and control, and for ensuring that appropriate corrective action is taken to address risks and issues. The status of the testing will be reported to the stakeholders in regular status reports.

### 5.6.1 Managing Issues, Change Requests, and Defects

The management of issues and change requests is a normal part of project management. An *issue* can arise at any time during the project (e.g. a supplier to the project may go out of business, an employee may resign, specialized hardware for testing may not arrive in time, and so on), and an issue refers to a problem that has occurred which may have a negative impact on the project. The severity of the issue is an indication of its impact on the project, and the project manager needs to manage it appropriately.

A *software defect* is a flaw in the software that causes it to produce an incorrect result, and it needs to be corrected by the developer and retested. The testers will identify defects during the various types of testing, and these are reported to the development team. The defect report should provide sufficient information to enable the developers to perform the necessary corrections, and the arrival rate of defects and the number of open defects provide an indication of the quality of the software. The management of defects is described in more detail in Sect. 7.4.

A *change request* is a stakeholder request for a change to the scope of the project, and it may arise at any time during the project. The impacts of the change request (e.g. technical, cost, and schedule impacts on development/testing) need to be carefully considered, as a change introduces new risks that may adversely affect cost, schedule, and quality. It is essential to understand these impacts to enable an informed decision on whether to authorize or reject the change request to be made. The project manager may directly approve small change requests, with the impacts of a larger change request considered by the project *change control board* (CCB).

The activities involved in managing issues and change requests are summarized below:

- Log issue or change request
- Assess impact
- Authorization (or rejection) of change request
- Implementation
- Verification
- Closure.

The management of change requests is discussed in more detail in Sect. 7.5.

## 5.7 Project Governance During Testing

The *project board*[5] (or steering group) is responsible for directing the project, and it is directly accountable for the success of the project. It consists of senior managers and staff in the organization who have the authority to make resources available, to remove roadblocks, and to get things done (Fig. 5.4).

It is consulted whenever key project decisions need to be made, and it plays a key role in project governance. The project board ensures that there is a clear business case for the project, and that the capital funding for the project is adequate and well spent. The project board may cancel the project at any stage during project execution should there cease to be a business case, or should project spending exceed tolerance and go out of control.[6]

The project manager reports to the project board and sends regular status reports to highlight progress made as well as the key project risks and issues. The project board meets at an appropriate frequency during the project (with extra sessions held should serious project issues arise).

The test manager will communicate the test status regularly to the project manager during the project, and the test status and the key test risks and issues will be discussed at the project board. The project manager attends the project board meeting and presents all key project information (including testing). There are several roles on the project board (an individual could perform more than one role), and their responsibilities are summarized in Table 5.5.

The project board will carefully consider the status of the project as well as the input from the project manager before deciding on the appropriate course of action (which could include the immediate termination of the project if there is no longer a business case for it).

## 5.8 Test Reporting

The frequency of test reporting is defined in the project test plan (or the communication plan). There is an IEEE standard (IEEE 829) for a test summary report. The test report advises management and the key stakeholders of the current status of the testing and includes key project testing information such as:

– Summary of testing activities and results
– Completed deliverables (during period)
– New risks and issues

[5]The project board in the PRINCE2 methodology includes roles such as the project executive, senior supplier, senior user, project assurance, and the project manager. These roles have distinct responsibilities.

[6]The project plan will usually specify a *tolerance level* for schedule and spending, where the project may spend (perhaps less than 10%) in excess of the allocated capital for the project before seeking authorization for further capital funding for the project.

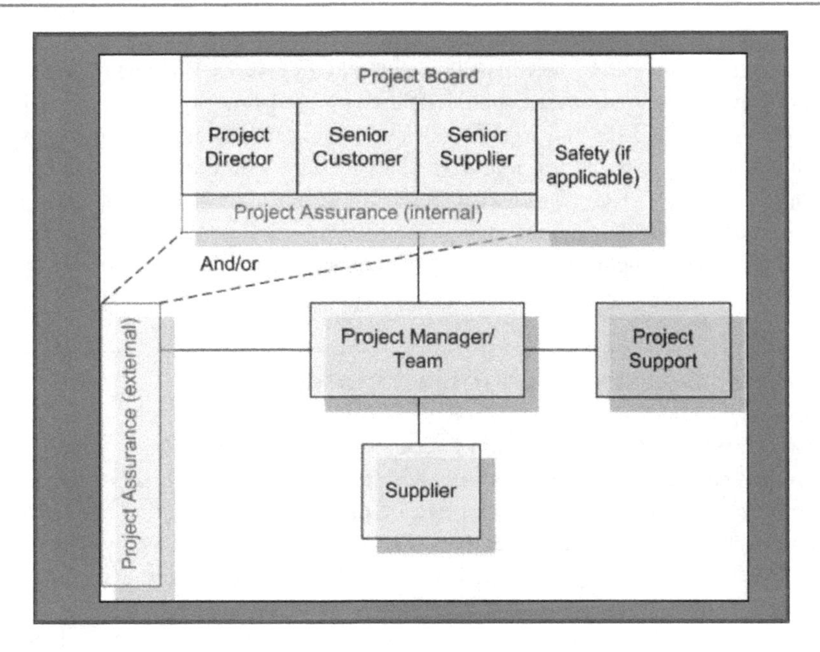


**Fig. 5.4** PRINCE2 project board

**Table 5.5** Project board roles and responsibilities

| Role | Responsibility |
|---|---|
| Project director | Ultimately responsible for the project. Provides overall guidance to the project |
| Senior customer | Represents the interests of users |
| Senior supplier | Represents the resources responsible for implementation of project (e.g. IS manager) |
| Project manager | Link between project board and project team |
| Project assurance | Internal role (optional) that provides an independent (of project manager) objective view of the project |
| Safety (optional) | Ensure adherence to health and safety standards |

– Schedule, effort, and budget status (e.g. RAG metrics[7])
– Test status
– Key risks and issues
– Milestone status
– Activities and deliverables planned (next period).

[7]Often, a colour coding mechanism is employed with a red flag indicating a serious issue; amber highlighting a potentially serious issue; and green indicating that everything is ok.

The test manager discusses the test report with management and presents the current status of the testing as well as the key risks and issues. The test manager will explain how the key issues are being dealt with and how the key risks will be managed. The new risks and issues will also be discussed, and management will carefully consider how the test manager plans to deal with these, and will provide appropriate support. The project manager will present a recovery plan (exception report) to deal with the situation where the project has fallen significantly outside the defined project tolerance (i.e. it is significantly behind schedule or over budget).

## 5.9 Lessons Learned and Project Closure

A project is a temporary activity, and once the project goals have been achieved and the product handed over to the customer and support group, it is ready to be closed. The project manager will prepare an end of project report detailing the extent to which the project achieved its objectives. The report will include a summary of key project metrics (including key quality metrics and the budget and timeliness metrics).

The success of the project is judged on the extent to which the defined objectives have been achieved and on the extent to which the project has delivered the agreed functionality on schedule, on budget, and with the right quality. This is often referred to as the project management triangle (Fig. 5.5).

The project manager presents the end project report to the project board, including any factors (e.g. change requests) that may have affected the timely delivery of the project or the allocated budget. The project is then officially closed.

The project manager and project team then consider the lessons learned during the projects, which are typically recorded in a lesson learned log. The key lessons learned are summarized in the lessons learned report, and the report is made available to other projects (with the goal of learning from experience). Any actions identified are assigned to individuals and followed through to closure. The project team is disbanded, and the project team members are assigned to other duties.

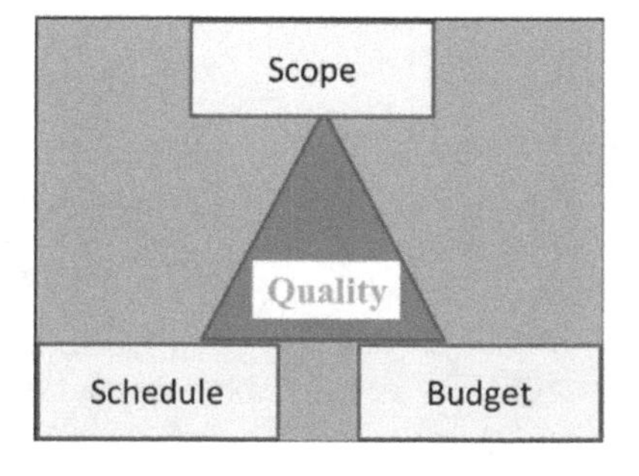


**Fig. 5.5** Project management triangle

## 5.10 Configuration Management

Configuration management is concerned with establishing and maintaining the integrity of the deliverables throughout the development lifecycle. It is concerned with:

– Configuration identification
– Configuration control
– Configuration control board (CCB)
– Baselining.

All test deliverables are uniquely identified and controlled. They are placed under configuration management control, including version control and change management. More detailed information on configuration management is in Chap. 15.

## 5.11 Review Questions

1. Describe the main activities in test planning?
2. Describe various approaches to estimation.
3. What skills are required to be a good test manager?
4. Explain the difference between the project test plan and the specific test plans for unit, system, and UAT.
5. What is the purpose of the project board?
6. What is the purpose of risk management? How are risks managed?
7. What is the difference between a risk and an issue?
8. How are defects handled during a project?
9. What is the purpose of test reporting?

## 5.12 Summary

Testing is a sub-project of a project and needs to be managed as such, and so good planning and monitoring and control are required. Test planning involves defining the scope of the testing to be performed; defining the test environment; estimating the effort required to define the test cases and to perform the testing; identifying the resources needed; assigning the resources to the tasks; defining the schedule; and identifying any risks to the testing and managing them.

The project test plan is developed and approved by the stakeholders, and maintained during the project. Estimation and scheduling are difficult as software projects are often complex and quite different from previous projects. Gantt charts are often employed for scheduling, and these show the work breakdown for the project, as well as task dependencies and the assignment of staff to the various tasks.

The effective management of risk is essential to project success. Risks arise due to uncertainty, and the risk management cycle involves risk identification; risk analysis and evaluation; identifying responses to risks; selecting and planning a response to the risk; and risk monitoring.

Once the test planning is complete, the focus moves to monitoring progress, managing risks and issues, re-planning as appropriate, providing regular progress reports to the project board, and so on. Finally, there is an orderly close of the project.

## Reference

Office of Government Commerce (2004) Managing successful projects with PRINCE 2. The Stationary Office, London

# Test Case Analysis and Design 6

**Key Topics**

Functional and non-functional requirements
Requirement traceability
Black box testing
White box testing
Statement coverage
Branch coverage
Equivalence partitioning
Boundary value analysis
Experienced-based testing
Decision tables
State transition testing
Use-case testing

## 6.1 Introduction

Test case analysis and design are concerned with analysing the requirements to determine the test conditions and designing the test cases (using various techniques). The requirements and test conditions are used to specify the test cases, where each test case includes input, a procedure for carrying out the test, and the expected results. The quality of the testing is influenced by the quality of the test cases, and they need to be designed to cover the requirements. Traceability of the



G. O'Regan, *Concise Guide to Software Testing*,
Undergraduate Topics in Computer Science,
https://doi.org/10.1007/978-3-030-28494-7_6

test cases to the requirements ensures that the test cases are sufficient to verify that all of the requirements have been implemented and tested.

The user requirements specify what the customer wants and define *what* the software system is required to do, as distinct from *how* this is to be done. They are determined from discussions with the customer to determine their actual needs, and they are then refined into the *system requirements*, which state the *functional* and *non-functional* requirements of the system. The requirements must be precise and unambiguous to ensure that all stakeholders are clear on what is (and what is not) to be delivered, and prototyping may be employed to clarify the requirements and to assist in their definition.

*Requirement verification* is concerned with ensuring that the requirements are properly implemented (i.e. *building it right*). In other words, it is concerned with ensuring that the requirements are properly addressed in the design and implementation, and a traceability matrix and testing are often employed as part of the verification activities.

Requirement validation (i.e. *building the right system*) is concerned with ensuring that the right requirements are defined, and that they are precise, complete, consistent, and realizable and reflect the actual needs of the customer. The validation is done by the stakeholders, and it may include several reviews of the requirements (and prototype), reviews of the design, and user acceptance testing.

The software design of the system is a blueprint of the solution of the system to be developed. It is concerned with the high-level architecture of the system, as well as the detailed design that describes the algorithms and functionality of the individual programs. The system architecture may include hardware such as personal computers and servers, as well as the definition of the subsystems with the various software modules and their interfaces. The choice of the architecture of the system is a key design decision, as it affects the performance and maintainability of the system.

The role of software testing is to reduce the risk of defects being present in the software and to increase confidence in its correctness. Test case design is concerned with the design and specification of the test cases to verify and validate the requirements and design. The traceability matrix is an effective way of verifying that the test cases cover all of the requirements and design. It involves mapping the requirements and design to the unit test cases; the system test cases; and the UAT test cases, and the traceability matrix provides a crisp summary of how the requirements have been implemented and tested.

Several types of testing are performed during the project, including unit, integration, system, regression, performance, and user acceptance testing. The software developers perform the unit testing, and the objective is to verify the correctness of a module. This type of testing is termed *"white box"* testing (also known as structural testing) and is based on knowledge of the internals of the software module. White box testing typically involves checking that every path in a module has been tested, and it involves defining and executing test cases to ensure code and branch coverage.

The objective of *"black box"* testing (also called specification-based testing) is to verify the functionality of a module (or feature or the complete system itself), and knowledge of the internals of the software module is not required. There are several specification-based techniques, which are used including use-case testing, equivalence partitioning, boundary value analysis, decision tables, and state transition testing.

## 6.2 Requirement Engineering

The process of determining the requirements for a proposed system involves discussions with the relevant stakeholders to determine their needs and to explicitly define what functionality the system should provide, as well as any hardware and performance constraints.

The specification of the requirements needs to be precise and unambiguous to ensure that all parties involved share a common understanding of the system and fully agree on what is to be developed and tested. A feasibility study may be needed to demonstrate that the requirements are feasible, and may be implemented within the defined schedule and cost constraints.

The requirements are the foundation for the system, and it is therefore essential that the requirements are *complete* (all services required by the user are defined), *consistent* (requirements should not contradict one another), and *unambiguous* (the requirements are clear and definite in meaning).

*Prototyping* may be employed to assist in the definition and validation of the requirements, and a suitable prototype will include key parts of the system. It will allow users to give early feedback on the proposed system and on the extent to which it meets their needs. Prototyping is useful in clarifying the requirements and helps to reduce the risk of implementing the incorrect solution.

We distinguish between the user (or business) requirements and the system requirements. The *user requirements* are the high-level requirements for the system (they tend to be high-level statements in a natural language with diagrams and tables), whereas the *system requirements* are a more detailed description of what the system is to do.

The system requirements include the functional and non-functional requirements. A *functional requirement* is a statement about the functionality of the system: i.e. a description of the behaviour of the system and how it should respond to particular inputs. A *non-functional requirement* is a constraint on the functionality of the system (e.g. a timing, performance, reliability, availability, portability, usability, safety, security, dependability, or a hardware constraint).

It is essential that the functional and non-functional requirements are stated precisely, and the *non-functional requirements are often quantitatively specified* so that it may be objectively determined (by testing) whether they are satisfied or not. Further, it is essential that the non-functional requirements are satisfied, as otherwise the delivered system may be unusable or unacceptable to the client. The non-functional requirements often affect the overall architecture of the system, rather than the individual components of the system.

## 6.3 Test Case Design Techniques

Test design involves defining the test conditions and test cases from the requirements and design documents, and establishing traceability to them to ensure that the test conditions/test cases are sufficient to verify the requirements/design. Each test case includes a test objective, input to the test case, the test procedure, and the expected results (Fig. 6.1).

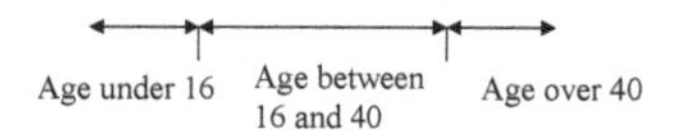


The test objective may be to find defects, obtain information on the quality of the software, provide confidence in the quality of the software, or prevent defects. The requirements/test conditions are the starting point for the specification of the test case, and the inputs are then identified. The test procedure is sufficiently detailed so that no expertise is required for execution of the test case, and test repeatability is guaranteed.

The formality of the testing depends on several factors including the maturity of the organization, the context of testing, the time constraints, and personnel involved. It is important that the testing is independent of the software development group. There are several test design techniques used for white box/black box testing (Table 6.1), which are discussed in more detail later in the chapter.

### 6.3.1 Black Box Testing

The objective of black box testing (also called *specification-based testing*) is to verify that the functionality of a module (or feature or the complete system itself)

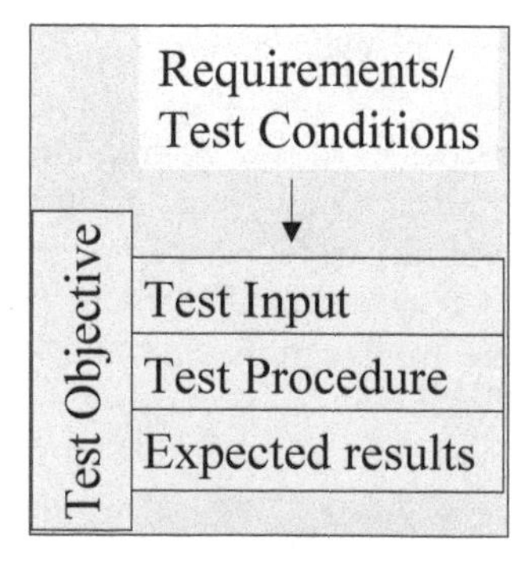


**Fig. 6.1** Test case specification

**Table 6.1** Test design techniques

| Type | Sub-type | Description |
|---|---|---|
| Black box testing (specification-based testing) | *Overview*<br>Equivalence partitioning<br>Boundary value Analysis<br>Decision table<br>State transition<br>Use-case testing | Based on functional and non-functional requirements<br>Use of equivalence classes for similar behaviour to reduce the number of test cases<br>This is based on the fact that many defects occur on the borders (boundaries)<br>Determine causes and effects; create decision table and test cases<br>Determine the states and events and valid and invalid transitions. Prepare test cases to verify correct behaviour at transitions<br>Determine users and their interactions (scenarios). Test cases are based on users and their interactions |
| White box testing (structure-based testing) | *Overview*<br>Statement testing<br>Decision testing | Test conditions/test cases are based on an analysis of the internal structure of the software. It is an indication of test coverage<br>Goal is to achieve 100% statement coverage in testing<br>Decision coverage is the percentage of decision outcomes that have been exercised in testing |
| Experienced-based testing | *Overview*<br>Error guessing<br>Exploratory testing | Testing is based on knowledge and experience of the tester<br>Experience/knowledge of tester is used to anticipate what defects might be present in the software<br>Tests are designed and executed at the same time (sometimes, they may not be recorded) |

satisfies the requirements, and knowledge of the internals of the software module is not required. There are several popular specification-based techniques such as equivalence partitioning, boundary value analysis, decision tables, state transition testing, and use-case testing.

**Equivalence Partitioning**

Equivalence partitioning involves dividing the test input data into a set of classes and selecting one input value from each class. The input selected is representative of the class, and this approach reduces an extremely large number of test cases (if all input data is used) to a manageable number of effective test cases.

For example, consider an application program that accepts an input range from 1 to 100. Then, this leads to three equivalence classes with one for valid input between 1 and 100, and two classes for invalid input (one for any data value below 1 and one for any value above 100).

**Boundary Value Analysis**

Boundary value analysis is an extension to equivalence partitioning and is based on the fact that many defects occur on the boundaries of the input domain (lower/upper values), rather than in the centre of the data range. Each boundary has a valid boundary value and an invalid boundary value, and the test cases are based on both valid and invalid boundary values.

For each input that concerns a range of values, the test cases are based on the boundaries and values just outside the boundaries. For example, the exact boundary values of the input range 1–100 are 1 and 100, and values just outside the boundaries are 0, 2, 99, and 101. For a range $\{r_1 \ldots r_2\}$, the exact boundary values are $r_1$, $r_2$, and the values just outside the boundaries are $r_1 - 1$, $r_1 + 1$, $r_2 - 1$, and $r_2 + 1$.

**Decision Tables**

A decision table (sometimes referred to as a cause–effect table) is a systematic way of stating business rules, and it provides a useful way to deal with a combination of different inputs and states. It is useful for testers when exploring combinations of different inputs and states that must correctly implement the business rules.

The use of a decision table in test design involves studying the specification, determining causes (including triggers) and effects (including actions), creating the decision table, and defining the test cases. The causes are required to be single conditions only with compound conditions split up, and the conditions are formulated in a positive way. All the combinations of true and false for the conditions are listed (if there are $n$ conditions, then since there are two choices for each condition, there are $2^n$ combinations for $n$ of them), and the corresponding action(s) to be carried out for that combination defined. A combination is sometimes referred to as a rule.

The analysis of the combinations may lead to combinations that were not mentioned in the original specification, and this is generally interpreted as that an error message should be displayed (i.e. another action may need to be added to display an error message). Further, the analysis of the combinations is useful in identifying omissions or ambiguities in the specification.

For example, consider a credit card system where there are three conditions on the discount received depending on whether you are a new customer, an existing customer with a loyalty card, or if you have a coupon. A new customer will receive a discount of 12% for today only, whereas an existing customer with a loyalty card will get a discount of 10%, and a person who has a coupon receives a discount of 15% today (but this offer cannot be used with the new customer discount).

The rules defining the discounts available today are summarized in the decision table in Table 6.2.

Each column of the decision table is formed from a combination of the conditions and represents the business rules (R1–R8). The approach to testing is generally to test each rule separately (if dealing with a small number of combinations),

**Table 6.2** Decision table with business rules

| Conditions | R1 | R2 | R3 | R4 | R5 | R6 | R7 | R8 |
|---|---|---|---|---|---|---|---|---|
| New customer | T | T | T | T | F | F | F | F |
| Existing/loyalty | T | T | F | F | T | T | F | F |
| Coupon | T | F | T | F | T | F | T | F |
| **Actions** | | | | | | | | |
| Discount (%) | X* | X* | 15 | 12 | 25 | 10 | 15 | 0 |

*X means that this is an invalid combination

and if there are a large number of combinations then a representative sample is chosen (this may involve prioritization of the combinations). There would generally be one test for each rule of the decision table, and this helps in finding defects.

The combinations of rules R1 and R2 should not arise, and an X for the discount indicates this (i.e. you cannot both be a new customer and have a loyalty card, and so there should be an error message). The discount for R3 is 15% [i.e. the maximum of the new customer discount (12%) and the coupon discount (15%)]. The discount for R5 is 25% [i.e. the sum of the existing customer discount (10%) and the coupon discount (15%)]. Rules R4, R6, and R7 have only one type of discount, and rule R8 had no discount (i.e. 0%).

**State Transition Testing**

State transition testing is employed when part of the system may be described with a finite state machine. That is, the system may be in a finite number of states, and the transitions from one state to another are determined by the rules of the machine. A state transition model has several parts including:

– States of the system
– Transitions from one state to another
– Events that cause a transaction
– Actions that result from a transition.

A state diagram is used to illustrate the finite state machine model of the system, where the states are shown as circles, the transitions are shown as arrows between circles, and events are shown as text near the transitions. State transition testing is useful at identifying defects at transitions and in identifying behaviour when none should be present.

The state diagram in Fig. 6.2 is a simplified model of PIN authentication for an ATM. It has five states and four possible events (Card Inserted, Enter PIN, Invalid PIN, and Valid PIN).

The test cases are then derived from typical scenarios such as the normal situation where the PIN is entered correctly, and then a second scenario where an incorrect PIN is entered. Test conditions may be derived from a state graph in various ways, as each state and each transition may be treated as a test condition.

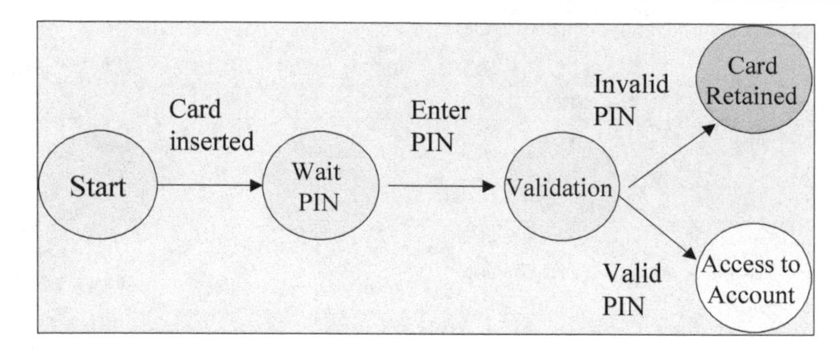


**Fig. 6.2** State diagram for PIN authentication

A state transition table shows the state that a finite state machine will move to from its current state and the input. It is useful in designing test cases for *n*-switch coverage, where the state transition table consists of the start state, the inputs, the expected output, and end state. The testing of all valid transitions is known as "0-switch" coverage. It is also possible to consider transition pairs (1-switch coverage) or triples (2-switch coverage), as well as coverage of *n* transitions (*n*-1-switch coverage).

**Use-Case Testing**

A use-case diagram models the dynamic aspects of the system, and it shows a set of use cases and actors and their relationships. It describes scenarios (or sequences of actions) in the system from the user's viewpoint (actor) and shows how the actor interacts with the system. An actor represents the set of roles that a user can play, and the actor may be human or an automated system.

Use-case testing is a form of testing that identifies test cases to exercise the whole system based on the interactions that a user has with the system. Each use case has a basic path/flow as well as failure scenarios.

A use-case diagram shows a set of use cases, with each use case representing a functional requirement. Use cases are employed to model the visible services that the system provides within the context of its environment, and for specifying the requirements of the system as a black box. Each use case carries out some work that is of value to the actor, and the behaviour of the use case is described by the flow of events in text. The description includes the main flow of events for the use case and the exceptional flow of events.

Use cases provide a way for the end users and developers to share a common understanding of the system (Jacobson et al. 1999). They may be applied to all or part of the system (subsystem), and the use cases are the basis for development and testing. The use cases describe the process flow through a system based on the likely use of the system, and the test cases derived from the use cases are very useful in finding real-world defects that users are most likely to find.

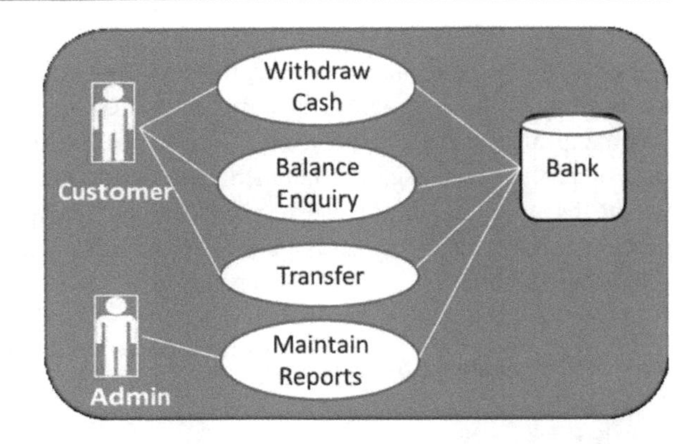


**Fig. 6.3** Use-case diagram of ATM

An ellipse is used to represent a use case graphically, and Fig. 6.3 presents a simple example of use cases in an ATM application. The typical user operations at an ATM include the balance inquiry operation, cash withdrawal, and the transfer of funds from one account to another. The actors for the system include "customer" and "admin", and these actors have different needs and expectations of the system.

The behaviour from the user's viewpoint is described, and the use cases include "withdraw cash", "balance enquiry", and "transfer", whereas the behaviour for the admin actor is to "maintain/reports". The use-case view includes the actors who are performing the sequence of actions.

Each use case must specify the preconditions that must be satisfied for the use case to work and the post-conditions that must be satisfied following the execution of the use case. Each use case has a most likely scenario and possibly additional scenarios covering special cases or exceptional conditions. Use case testing involves testing the main scenario and one for each alternate scenario.

The advantages of use-case testing are that the main tasks/functions of the system are being tested, and that the testing is user-focused and designed to find real-world defects.

### 6.3.2 White Box Testing

White Box Testing (also called *structure-based techniques*) involves the design of test conditions and test cases based on an analysis of the internal structure of the software. It may be used to measure the coverage of existing test cases as well as to derive further test cases to increase coverage. We distinguish between *statement coverage* (where 100% statement coverage means that every statement is executed at least once) and *decision coverage* (*branch coverage*), which has the additional requirement that all possible outcomes of every decision will be tested. There are several types of white box testing including statement testing and decision testing.

**Statement Testing**

A statement is a syntactic entity in a programming language that expresses an action to be performed. A program is a collection of one or more statements, and a statement may include expressions. The objective of statement coverage is to determine the percentage of statements executed during testing and to determine where the code has not been executed due to blocked tests. The objective is to process and execute every line of code, and the statement coverage is given by the following formula:

$$\text{Statement Coverage} = \frac{\text{Number Statements Executed}}{\text{Total Number of Statements}} \times 100\%$$

Statement coverage tests cannot test false conditions in a statement (i.e. if the condition part of an if statement is false, then the statement is not executed). For example, consider the following code fragments written in a C like syntax.

| Fragment A | Fragment B |
|---|---|
| 1. x = 2; | 1. x = 3; |
| 2. y = 3; | 2. y = 2; |
| 3. if (x > y) | 3. if (x > y) |
| 4. x = x + 1 | 4. x = x + 1 |

There are four statements in each fragment. However, three statements are executed for fragment one (i.e. the statement coverage is 75%), whereas four statements are executed for fragment two (i.e. the statement coverage is 100%).

**Decision Testing**

Decision testing (also called branch testing) covers both the true and the false conditions as distinct from statement testing, which covers only true conditions. A decision is a point in the program where the control flow has two or more alternative routes, and decision coverage (branch coverage) is the percentage of decision outcomes that have been executed by the suite of tests.

$$\text{Decision Coverage} = \frac{\text{Number Decision Outcomes executed}}{\text{Total Number of Decision Outcomes}} \times 100\%$$

A branch is the outcome of a decision, and branch coverage measures the extent to which decision outcomes have been tested. It is applicable to statements such as *if statements*, *while statements*, *repeat statements*, and *case statements*. The branch coverage metric is given by the formula above. Decision testing measures the extent to which all branches in the software code have been reached, and 100% decision coverage guarantees 100% statement coverage (but not the other way around). The following code fragment illustrates decision testing.

```
1. if (x > y)
2.      x = x + 1
3. else
4.    x = x − 1
```

Then, we need two test cases to achieve 100% decision coverage, and the following two test cases will suffice:

<u>**Test Sets**</u>

T1: x is 3 and y is 1
T2: x is 3 and y is 4

### 6.3.3 Experienced-Based Testing

The approach in experienced-based testing is to use the knowledge and experience of the tester to prepare test conditions and test cases. The personnel involved have previous experience of similar projects and are familiar with the software and its environment. They often have an insight into what could go wrong with the software, which is potentially very useful in identifying defects during testing.

These techniques are generally used after the completion of black box and white box testing, rather than as a replacement for these activities. The two main experienced-based techniques are error guessing and exploratory testing.

**Error Guessing**
This approach uses the experience of the tester to anticipate which defects might be present in the software. Often, the form of testing proceeds without any formal test documentation and relies on the knowledge and expertise of the tester. The success of this type of testing is dependent on the expertise and skills of the tester, as experienced testers will know where the defects are most likely to be. However, the disadvantage with error testing is that it is not repeatable as such, and it is random in the sense that it is trying things out rather than the normal planned activities.

**Exploratory Testing**
Exploratory testing is concerned with exploring the software and determining what works and does not work. The tester decides what to test next and where to spend the limited amount of time available, and tests are often designed and executed at the same time. The objective is to spend the minimal time in planning and the maximum time in execution.

Test design and execution activities are performed in parallel without formally documenting the test conditions and test cases. This type of testing is often performed after the more formal testing has been completed as a check to ensure that the most serious defects have been identified in the software.

## 6.4 Test Case Specification

Several types of testing that may be performed were described in Table 3.1, and there is often a separate test plan for unit, system, and UAT. The unit tests are based on the software design, the system tests are based on the system requirements, and the UAT tests are based on the user requirements. The dedicated test plans generally include a planning section as well as the test scripts for the testing (Tables 6.3 and 6.4).

Each of these test plans contains test scripts (e.g. the unit test plan contains the unit test scripts and so on), and the test scripts are traceable to the design (for the unit tests) and for the system requirements (for the system test scripts). The unit tests are more focused on white box testing, whereas the system test and UAT tests are focused on black box testing.

Each test script contains an objective(s) and the procedure by which the test is carried out. A test script generally includes:

- Test case ID
- Test type (e.g. unit, system, UAT)
- Objective(s)/description
- Test script steps (for each objective)
- Expected results
- Actual results
- Tested by.

Regression testing involves carrying out a subset of the defined tests to verify that the core functionality of the software remains in place following changes to the system (e.g. correction of defects or addition of new functionality).

The test plans are often described in a test document, but they could also be defined using a test management tool (see Chap. 10 for a discussion of tools for software testing). The dedicated test plan will detail the planning specific to the type of testing being conducted including the testing to be performed, the approach, the resources and training required, the planned preparation dates, the test environment

**Table 6.3** Planning section in dedicated test plan

| | |
|---|---|
| Description (including objectives) | |
| Approach | |
| Resources and responsibilities | |
| Training required | |
| Preparation dates | |
| Testing dates | |
| Test environment | |
| Testing tools | |
| Entry criteria | |
| Exit criteria | |
| Features to test | |

**Table 6.4** Template for test case

| **TEST CASE TEMPLATE** | | | | |
|---|---|---|---|---|
| **Title** | | | **Req ID** | |
| **Author** | | | **Test Case ID** | |
| **Date** | | | **Test Type** | |
| | | | | |
| **Objective** | **Objective Description** | | | |
| # | Enter context of Objective # here | | | |
| # | Enter context of Objective # here | | | |
| # | Enter context of Objective # here | | | |
| # | Enter context of Objective # here | | | |
| # | Enter context of Objective # here | | | |
| **Test Objective #** | Enter Test Objective | | | |
| **Test Input** | Enter Input for test | | | |
| **Test Script Step(s)** | Enter Test Script Step(s) here | | | |
| **Expected Results** | Enter Expected Results of test | | | |
| **Actual Results** | Enter Actual Results of test | | | |
| **Data Location** | Enter Location of Test Data here | | | |
| | **Name** | **Test Date** | **Test Status** | **Defect No.** |
| **Tested By** | | | | |
| **Reviewed By** | | | | |
| **Approved By** | | | | |
| | | | | |
| **Test Objective #** | Enter Test Objective | | | |
| **Test Input** | Enter input for test | | | |
| **Test Script Step(s)** | Enter Test Script Step(s) here | | | |
| **Expected Results** | Enter Expected Results of test | | | |
| **Actual Results** | Enter Actual Results of test | | | |
| **Data Location** | Enter Location of Test Data here | | | |
| | **Name** | **Test Date** | **Test Status** | **Defect No.** |
| **Tested By** | | | | |
| **Reviewed By** | | | | |
| **Approved By** | | | | |

and test tools required, and the entry and exit criteria. It will generally include risks and assumptions, as well as a traceability matrix that maps the test cases to the requirements or design.

The test plan will contain the test scripts to carry out the testing, where a test script defines the steps required to carry out a particular test (Table 6.4). Each test script is based on the test conditions that have previously been defined, and may contain several test objectives (or subtests). Each test objective includes a description, the inputs to the test case, the test procedure (or steps required to carry out the test), the expected output, the actual results, and whether the test passes or fails. The results of the test scripts will be recorded including details of who carried out the tests.

Table 3.1 describes the types of testing that may be carried out during the project such as unit testing, component testing, system testing, performance testing, load/stress testing, browser compatibility testing, usability testing, security testing, regression testing, test simulation, and acceptance testing.

## 6.5 Requirement Traceability

Requirement traceability was briefly discussed in Sect. 3.11 and provides a way to verify that all of the defined requirements for the project have been implemented and tested. One way to do this is to consider each requirement number and to go through every part of the design document to find where the requirement is being implemented in the design, and similarly to go through the test documents and find any reference to the requirement number to show where it is being tested. This would demonstrate that the particular requirement number has been implemented and tested.

A more effective way to do this is to employ a traceability matrix (Table 6.5), which may be employed to map the user requirements to the system requirements; the system requirements to the design; the design to the unit test cases; the system test cases; and the UAT test cases. The matrix provides a crisp summary of how the requirements have been implemented and tested.

The traceability of the requirements is *bidirectional*, and the traceability matrix may be maintained as a separate document or as part of the requirement document. The basic idea is that a mapping between the requirement numbers and the sections of the design or test plan is defined, and this provides confidence that all of the requirements have been implemented and tested.

Requirements will usually be numbered, and a single requirement number may map on to several sections of the design or to several test cases: i.e. the mapping may be *one to many*. The traceability matrix provides the mapping between the individual requirement numbers and the sections in the design or test plan corresponding to the particular requirement number.

**Table 6.5** Sample trace matrix

| Requirement No. | Sections in design | Test cases in test plan |
|---|---|---|
| Rl.1 | D1.4, D1.5, D3.2 | T1.2, T1.7 |
| R1.2 | D1.8, D8.3 | T1.4 |
| R1.3 | D2.2 | T1.3 |
| … | … | … |
| R1.50 | D20.1, D30.4 | T20.1, T24.2 |

It is essential to keep the traceability matrix up to date during the project and especially after changes to the requirements. The traceability matrix is useful in determining the impacts of a proposed change to the requirements, as it enables the impacts on other requirements and project deliverables to be easily determined.

## 6.6 Review Questions

1. What is the difference between a functional and non-functional requirement?
2. What is the difference between requirement verification and validation?
3. Explain the difference between black box testing and white box testing.
4. Describe the main specification-based techniques used in black box testing.
5. Describe the main structure-based techniques used in white box testing.
6. Describe the main experienced-based testing techniques.
7. What is the purpose of requirement traceability?
8. Explain the difference between statement coverage and branch coverage.
9. Explain use-case testing.
10. Explain equivalence partitioning and boundary value analysis.

## 6.7 Summary

Test analysis and design are concerned with analysing the requirements to determine the test conditions and designing the test cases (using various techniques) for the testing. The requirements and test conditions are used to specify the test cases, where each test case includes input, the procedure for carrying out the test, and the expected results. The quality of the testing is influenced by the quality of the test cases. Traceability of the test cases to the requirements is essential in ensuring that the testing is sufficient to verify that all of the requirements have been implemented.

The user requirements specify what the customer wants and define what the software system is required to do, as distinct from how this is to be done. The requirements are the foundation for the system, and the process of determining the requirements, analysing and validating them, and managing them throughout the project lifecycle is termed requirement engineering.

The objective of black box testing is to verify that the functionality of a module (or feature or the complete system itself) satisfies the requirements, and knowledge of the internals of the software module is not required. There are several popular specification-based techniques such as equivalence partitioning, boundary value analysis, decision tables, state transition testing, and use-case testing.

White box involves the design of test conditions and test cases based on an analysis of the internal structure of the software. It may be used to measure the coverage of existing test cases, and white box testing includes statement testing and decision testing.

The approach in experienced-based testing is to use the knowledge and experience of the tester to prepare test conditions and test cases. The personnel involved often have an insight into what could go wrong with the software, which is useful in identifying defects.

The objective of requirement traceability is to verify that all of the defined requirements for the project have been implemented and tested. The traceability matrix provides a crisp summary of how the requirements have been implemented and tested.

## Reference

Jacobson I, Booch G, Rumbaugh J (1999) The unified software modelling language user guide. Addison-Wesley, Boston

# Test Execution and Management 7

**Key Topics**

Test team
Test monitoring and control
Test execution
Change requests
Defect management
Risk management
Test reporting
Test completion criteria

## 7.1 Introduction

Test management is concerned with the activities involved in managing the software testing, whereas test execution is concerned with the activities involved during the execution of the test cases. The main activities in test management include the organization of the test team, test planning, test case design and specification, test execution, defect management, change request management, test monitoring and control, and test reporting. Good test management is a key enabler to project success, and an effective test process is repeatable and predictable.

Test management may involve the use of a dedicated test tool (e.g. HP Quality Center), which ensures that the testing follows the defined process with process discipline enforced by the tool. This is useful when there are regulatory requirements to be satisfied and where a full audit trail of the testing is required.


G. O'Regan, *Concise Guide to Software Testing*,
Undergraduate Topics in Computer Science,
https://doi.org/10.1007/978-3-030-28494-7_7

Several test tools may be employed during test execution such as capture/playback tools for regression testing, and automated tools to simulate a large number of users in performance testing.

The test team carries out the scheduled test activities such as setting up the test environment, defining the test cases, running the test cases, logging defects and communicating problems to the developers, and retesting failed tests. The test manager monitors progress during execution and takes corrective action to manage the risks and issues that arise to ensure that the testing remains on track. The test manager communicates progress regularly to management during the testing, and the project manager reports the key test status to the project board during the project. The main activities involved in test execution and management are summarized in Table 7.1.

**Table 7.1** Test management activities

| Activity | Description |
|---|---|
| Test team organization | The testing may be conducted by an internal test team or outsourced to an external company. The decision to outsource may be based on the availability of skilled personnel, cost, time pressures, and the known risks |
| Test planning | Test planning is concerned with the planning for the testing, including the scope of the testing, the activities to be performed, when they will be performed as well as who will perform them. There will usually be an overall test plan (the project test plan) and separate specialized test plans for the different types of testing performed |
| Test design and specification | This activity is concerned with writing the test cases and consists of defining the test conditions and specifying the test cases, as well as the steps to carry out the tests and the expected results. The test scripts may be manually created or entered into a test tool for automation |
| Test execution | This activity is concerned with the execution of the test scripts and comparing the actual results to the expected results to determine if the test passed or failed. The execution may be manual or automated with a test tool |
| Test monitoring and control | This is concerned with monitoring test execution and taking corrective action when performance deviates from expectations |
| Configuration management | Configuration management is concerned with establishing and maintaining the integrity of the test deliverables throughout the development lifecycle |
| Risk management | This is concerned with the identification and management of risk during the testing |
| Managing defects | Defect management is concerned with the logging and resolution of defects that are identified during test execution |
| Managing change requests | This is concerned with the management of change requests that arise during the project and verifying that they have been implemented correctly |
| Test reporting and metrics | Test reporting is concerned with communicating the results of the testing to the stakeholders. It involves preparing test metrics and reports |
| Test completion criteria | This specifies the criteria for determining when the planned testing is complete (it may be part of the release criteria) |

## 7.2 Test Planning

Test planning was discussed in Chap. 5, and the test manager will create the project test plan and schedule. The test manager will track the schedule to completion, and the schedule will detail the key test milestones, the activities and tasks to be performed as well as their associated timescales, and the resources required to carry out each task. The test manager will update the project schedule regularly during the project.

The project test plan is documented (this could be part of the project plan, but it is usually a separate document), and it defines how the testing will be carried out. It includes the scope of the testing, the organization of the test team and the personnel involved, the resources and effort required, the key milestones, the definition of the testing environment, any special hardware and test tools required, and the planned test schedule.

There is generally a separate test plan for the various types of testing (see Chap. 6), which records the planning for the particular type of testing as well as the test cases, including the purpose of each test case, the inputs and expected outputs, and the test procedure for the execution of the particular test case.

### 7.2.1 Test Team Organization

Test team management is concerned with organizing the test team to ensure that high-quality testing is completed on time and budget during the project. The organization of a simple test team is described in Fig. 7.1, and the test manager plans, monitors, and controls the testing. The test manager role was described in Chap. 5, and it involves defining and maintaining the test plan and schedule, as well as preparing and communicating the regular test reports during the project.

It is essential that sufficient time is allowed for testing, and that the required physical and human resources are provided. The test manager needs to be proactive in ensuring that the quality of the testing is not compromised should the project fall

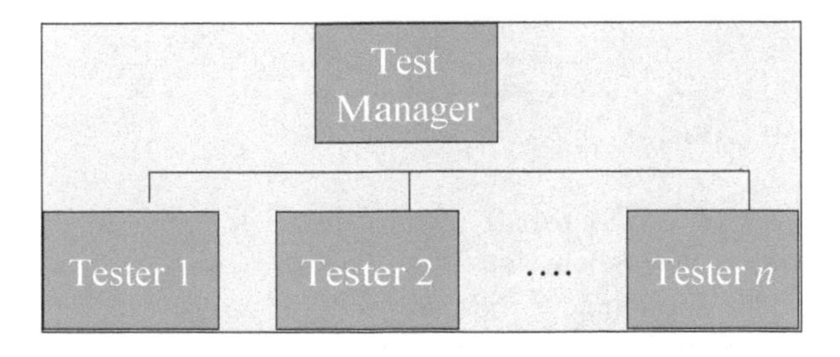


**Fig. 7.1** Organization of a test team

behind schedule. This is since project pressure could be applied to stay with the original schedule and timelines, with a corresponding reduction of the amount of time available for testing, and a subsequent negative impact on quality.

Test tools may simplify the management process of a test team, as they generally provide a structured approach to track test execution and management of distributed test teams. Further, it is essential that the testers are properly trained in testing, and that they are familiar with best practice in software testing.

The testers participate in the review of the requirements (from the test viewpoint) and specify the test cases to verify the requirements. They are responsible for execution of the test cases and recording the results, and for logging and communicating defects. They may participate in the automation of tests where automation tools are part of the process. Often, testers will specialize in a particular area.

It has become more common for testing to take place in different geographical locations, and so the management of distributed teams in different parts of the world has become an important consideration. Test outsourcing has become popular, and so the selection and management of an external test team have become important (see Chap. 8).

The testing will usually be led by a test leader/coordinator (or test manager) who is responsible for coordinating the test activities, assigning team members to the various activities, monitoring progress and managing schedule, managing risks and issues that arise during testing.

The test team will need to maintain a close relationship with both the business analysts (or requirement group) and the developers (software development organization). It is essential that the test cases cover the user requirements (i.e. they need to be traceable to the requirements), and that once the tests have been run that the results are analysed and any failed tests identified. The failed tests will generally result in defects that need to be reported to the developers for correction. The testers will need to be diplomatic in reporting defects to the developers, and sensitivity may be required in order to keep a good professional relationship.

Test quality and productivity may be improved with test automation, and it is very useful to replace highly repetitive tests such as regression testing with an automated tool.

## 7.3 Test Execution

The software developers will normally carry out the unit and integration testing as part of the software development activities. The developers will correct any identified defects, and the development continues until all unit and integration tests pass, and the software is fit to be released to the test group.

The software test group will conduct the test activities such as system testing, performance testing, usability testing, and so on. There will usually be a formal handover from development to the test group prior to the commencement of testing, and the *entry criteria* specify what needs to be satisfied in order for the software to be accepted for testing by the test group. The handover criteria may be similar to:

- All unit and integration tests have been run and passed.
- All known risks are identified.
- The test environment is ready for testing.
- All relevant test scripts (e.g. system and performance) are prepared and documented.
- All required resources are available.

Test execution then commences, and the testers run the system tests and other tests, log any defects in the defect tracking tool, discuss with the development team, and communicate progress to the test manager. The status of the testing is communicated regularly to the project team, and the developers correct the identified defects and produce new releases of the software. The test group retests the failed and carries out tests that were previously blocked.

Regression testing is performed to ensure that the core functionality of the system is preserved and that no new problems have been introduced following the changes to the software. This continues until the testing has been completed, and the *exit criteria* for test completion specify the criteria to be satisfied for testing to be considered complete. It may be similar to:

- All system, performance, and other tests are run and passed.
- All agreed defects are corrected.
- All risks are known and can be managed.

## 7.4 Managing Defects

The execution of the test cases by the testers may result in the situation where the actual results obtained differ from the expected results, and this generally results in a defect report[1] (or bug[2]) being generated (Fig. 7.2). It is important to log all problems that arise during testing, and this is often done with a defect tracking tool (it may involve a defect tracking spreadsheet for logging defects and a defect form for small projects).

Each defect (bug) should be described in detail including:

- Problem number
- Severity

[1]The difference between expected/actual results could be due to factors rather than a genuine defect such as poor test data, errors made by the tester, or invalid expected results.

[2]Grace Murray Hopper coined the term "computer bug" when she traced an error in the Harvard Mark II computer to a moth stuck in one of the relays. The bug was carefully removed and taped to a daily logbook, and the term is now ubiquitous in the computer field

**Fig. 7.2** Computer bug

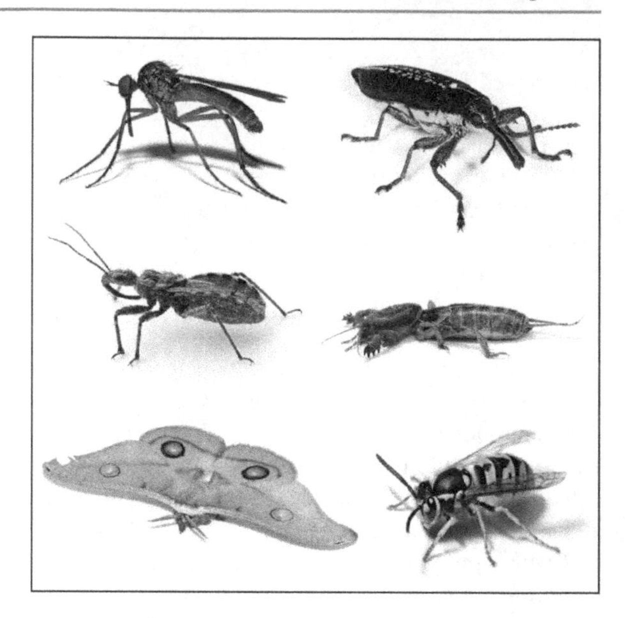

- Tester
- Date raised
- Status
- Description of problem and steps to reproduce
- Impacts of problem
- Responses to problem
- Implementation and verification.

The severity of the defect indicates how serious the problem is, and there are usually several categories of severity such as:

- Critical
- Major
- Medium
- Minor.

Defects are scheduled for correction in a later release of the software, and the correction involves analysis by the developers to determine the cause of the problem and appropriate updates to the software to resolve it. The updated software needs to be retested to ensure that the defects have been resolved, and that no new defects have been introduced. Regression testing is performed to ensure that the core functionality of the system is preserved.

The status of a defect indicates whether it has been resolved or not. It could be one of the following:

– Open (defect identified by test team)
– Assigned (defect assigned to developer for resolution)
– Changes implemented (developers have implemented changes)
– Changes verified (testing has been conducted to verify changes)
– Closed (defect closed).

## 7.5 Managing Change Requests

Change control is concerned with the formal management of change throughout the project. It is important that changes are controlled and that their impacts are clearly understood. For example, a change to the requirements generally affects other project deliverables such as the design, code modules, and test documents, and it is essential to keep change to a minimum towards the end of the project.

The project manager may authorize small change requests, but larger change requests need to be reviewed and authorized by the change control board (CCB). Changes to the requirements may introduce new risks to the project, and it is essential that these be considered by the CCB to ensure that they can be managed. The activities involved in formal change control are described in Table 7.2.

In other words, a formal request to change the requirements is logged, and the change control board assesses its impacts. The CCB decides on whether to authorize or reject the proposed change, and if authorization is given the developers will implement the solution and the testers will verify its correctness and that no new defects have been introduced.

**Table 7.2** Activities in managing change requests

| Activity | Description of change request |
|---|---|
| Log change request | The project manager logs the change request. It is assigned a unique reference number and priority (severity) |
| Assess impact | This involves analysis to determine the impacts such as technical, cost, schedule, risks, and quality |
| Decision on implementation | A decision is made on how to deal with the change request is generally made by the CCB |
| Implement solution | The affected documents and software modules are identified and modified accordingly |
| Verify solution | Testing (unit, system, and UAT) is employed to verify the correctness of the solution |
| Close issue/CR | The issue or change request is closed |

## 7.6 Test Monitoring and Control

Test monitoring and control are concerned with monitoring test execution and taking corrective action when performance deviates from expectations (Fig. 7.3). The progress of the testing and key milestones should be monitored against the plan, and corrective actions taken as appropriate. The key parameters such as budget, effort, and schedule as well as risks and issues are monitored, and the status of the testing communicated regularly to the affected stakeholders.

The test manager will conduct progress and milestone reviews to determine the actual progress, with new issues identified and monitored. The appropriate corrective actions are identified and are tracked to closure. The main focus of test monitoring and control is:

- Monitor the test plan and schedule and keep on track.
- Monitor the key project parameters.
- Conduct progress and milestone reviews to determine the actual status.
- Re-plan as appropriate.
- Monitor risks and take appropriate action.
- Analyse issues and change requests and take appropriate action.
- Track corrective action to closure.
- Monitor resources and manage any resource issues.
- Report the test status to management.

Figure 7.3 presents a sample process map for test monitoring and control.

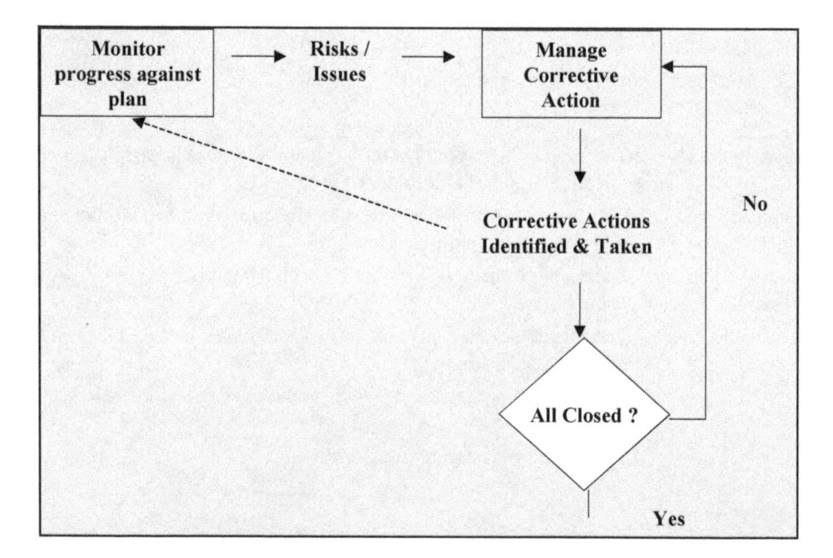


**Fig. 7.3** Test monitoring and control process map

## 7.7 Risk Management

Risk management was discussed in Chap. 5, risks arise due to uncertainty, and risk management is concerned with managing uncertainty and especially the management of any undesired events. Risks need to be identified, analysed, and controlled in order for the testing to be successful, and the risk management activities take place throughout the project lifecycle.

The test manager is responsible for identifying the initial set of risks to testing, and they are analysed to determine their likelihood of occurrence and their impact (e.g. on cost, schedule, or quality) should they occur.

Countermeasures are defined to reduce the likelihood of occurrence and impact of the risks, and contingency plans are prepared to deal with the situation of the risk actually occurring. Additional risks may arise during the testing, and the test manager needs to be proactive in their identification and management.

Risks need to be reviewed regularly, especially following changes to the project. These could be changes to the business case or the business requirements, loss of key personnel, and so on. Events that occur may affect existing risks (including the probability of their occurrence and their impact) and may lead to new risks. Countermeasures need to be kept up to date during the project. Risks are reported regularly throughout the testing, and the risk management activities include identifying risks; determining their probability of occurrence and impact should they occur; identifying responses to the risks; and monitoring and reporting.

The test manager will maintain a risk repository (this may be a part of the project risk repository such as a tool or a risk log) to record the details of each risk, including its type and description; its likelihood; its impact; as well as the appropriate response to the risk.

## 7.8 Test Reporting

Test reporting is concerned with communicating the status of the testing to the stakeholders during the project. The test manager will prepare regular test reports for the key stakeholders during the project, and the report summarizes the testing that has taken place during the period. The test report template may be based on the IEEE 829 standard for a test summary report, and it will generally include key project information such as:

- Completed test deliverables (during period)
- New risks and issues
- Schedule, effort, and budget status for testing (e.g. usually metrics[3])
- Test status

[3]For example, the budget estimation accuracy metric is given by (actual spend—original budget)/ original budget * 100 (with similar formulae for schedule and effort estimation accuracy)

– Quality status
– Key risks and issues
– Milestone status
– Activities and deliverables planned (next period).

The test manager discusses the test report with management and presents the current status of the testing as well as the key risks and issues. The test manager works closely with the project manager, and they will present a recovery plan (exception report) to deal with the situation where the project/testing has fallen outside the defined project tolerance (i.e. it is significantly behind schedule or over budget).

The overall test status for the project could be presented in a way similar to Table 7.3. This provides a crisp summary of the testing that has been carried out as well as what needs to be done.

The quality status of the project is given by a crisp summary of the number of open defects by severity and could be described in a way similar to Table 7.4. The key risks and issues affecting testing will be discussed, and the test manager will explain how these are being dealt with (Table 7.5). The new risks and issues will also be discussed, and the project board will carefully consider how the test manager plans to deal with these, and will provide appropriate support.

**Table 7.3** Test status for project

| Test type | # Scripts | # Run | # Pass | # Fail | % Run | % Pass |
|---|---|---|---|---|---|---|
| Unit | 50 | 50 | 50 | 0 | 100 | 100 |
| System | 100 | 80 | 72 | 8 | 80 | 90 |
| Regression | 50 | 50 | 50 | 0 | 100 | 100 |
| UAT | 20 | | | | | |
| Other | 10 | | | | | |

**Table 7.4** Quality status for project

| Severity | Total no. of defects | No. of open defects | Total no. of change requests | No. of open change requests |
|---|---|---|---|---|
| Sev 1 | 3 | 1 | | |
| Sev 2 | 10 | 2 | 2 | 0 |
| Sev 3 | 20 | 4 | | |
| Sev 4 | 15 | 7 | | |

**Table 7.5** Key risks for project key risks

| Risk no. | Description | Countermeasure |
|---|---|---|
| 1. | | |
| 2. | | |
| 3. | | |

The project board will carefully consider the status of the project as well as the input from the project/test manager before deciding on the appropriate course of action (which could include the immediate termination of the project if there is no longer a business case for it).

## 7.9 Test Completion Criteria

The test completion criteria specify the exit criteria to be satisfied for the testing to be considered complete. The testing may halt once the test completion criteria have been achieved as the software has achieved the desired quality criteria. The test completion criteria may be similar to:

- All system tests are run and passed.
- All performance tests are run and passed.
- All other tests are run and passed.
- All serious defects are corrected.
- Only minor defects open.
- Test report is complete and communicated.
- All risks can be managed.

## 7.10 Review Questions

1. Describe the activities in test planning.
2. Explain the difference between a change request and a defect.
3. What is the purpose of entry criteria and exit criteria?
4. Describe how defects are managed.
5. Describe how change requests are managed.
6. Describe the activities in test monitoring and control.
7. Explain how the quality of testing may become compromised if the project falls behind schedule.
8. Describe the activities in test reporting.
9. Explain the purpose of test completion criteria.
10. What is the purpose of risk management?

## 7.11 Summary

A well-planned and managed testing process enables teams to deliver high-quality products on time and on budget. Test management is concerned with the activities involved in managing the software testing, and good test management is a key enabler to project success. An effective test process is repeatable and predictable, and makes it easier to estimate and to plan and execute the testing.

The main activities involved in test management and execution include the organization of the test team, test planning, test case design and specification, test execution, defect management, change request management, test monitoring and control, and test reporting. Test management may involve the use of a dedicated test tool to ensure that the testing follows the defined process, and that process discipline is enforced. Several test tools may be employed during test execution such as tools for performance or regression testing.

The test team carries out the scheduled test activities such as setting up the test environment, defining the test cases, running the test cases, logging defects and communicating problems to the developers, and retesting failed tests. The test manager monitors the testing during execution and takes corrective action to manage the risks and issues that arise to ensure that the testing remains on track. The test manager communicates progress regularly to management during the testing, and the project manager reports the key test status to the project board during the project.

# Test Outsourcing

# 8

**Key Topics**

Request for proposal
Supplier evaluation
Formal agreement
Statement of work
Managing supplier
Service level agreement
Escrow
Acceptance of software

## 8.1 Introduction

Test outsourcing is concerned with the challenges of outsourcing the testing part of a project to a third-party testing organization. It is concerned with the selection of an appropriate supplier to perform the testing and the management of the supplier during the project. A project may lack the in-house expertise or resources to conduct the testing, and in such situations, it may be appropriate to outsource the testing to a specialized test organization. It is essential that the selected test organization is capable of carrying out the testing to the desired quality standard, as well as being capable of completing the testing within the budget and schedule constraints.

This means that the process for the selection of the supplier needs to be rigorous, and that the capability of the suppliers and their associated risks are known prior to selection. Supplier selection is generally based on objective criteria such as cost, the



G. O'Regan, *Concise Guide to Software Testing*,
Undergraduate Topics in Computer Science,
https://doi.org/10.1007/978-3-030-28494-7_8

approach, the ability of the supplier to deliver the required solution, the supplier capability, and while cost is an important criterion, it is just one among several other important factors.

Once the selection of the supplier is finalized, a legal agreement is drawn up between the contractor and supplier, which states the terms and condition of the contract, as well as the statement of work. The statement of work details the work to be carried out, the deliverables to be produced, when they will be produced, the personnel involved their roles and responsibilities, any training to be provided, and the standards to be followed.

The supplier then commences the defined work and is appropriately managed for the duration of the contract. This will involve regular progress reviews of the testing conducted, and acceptance testing is carried out prior to confirming that the testing is complete. Table 8.1 describes the activities generally employed for supplier selection and management.

**Table 8.1** Supplier selection and management

| Activity | Description |
|---|---|
| Planning and requirements | This involves defining the approach to the outsourcing of the testing. It involves<br>– Defining the procurement requirements for the testing<br>– Forming the evaluation team to rate each supplier against objective criteria |
| Identify suppliers[a] | This involves identifying suppliers capable of performing the required testing and may involve research, recommendations from colleagues or previous working relationships. Usually three to five potential suppliers will be identified |
| Prepare and issue RFP | This involves the preparation and issuing of the request for proposal (RFP) to potential suppliers. The RFP may include the evaluation criteria and a preliminary legal agreement |
| Evaluate proposals | The received proposals are evaluated and a shortlist produced. The shortlisted suppliers are invited to make a presentation of their proposed solution |
| Select supplier | Each supplier makes a presentation followed by a Q&A session. The evaluation criteria are completed for each supplier and reference sites checked (as appropriate). The decision on the preferred supplier is made |
| Define supplier agreement | A formal agreement is made with the preferred supplier. This may include<br>– Negotiations with the supplier/involvement with Legal Department<br>– Agreement may vary (statement of work, service level agreement, escrow, etc.)<br>– Formal agreement signed by both parties<br>– Unsuccessful parties informed<br>– Purchase order raised |
| Managing the supplier | This is concerned with monitoring progress with the testing carried out by the supplier, and managing risks, milestones and issues, and taking corrective action when progress deviates from expectations |
| Acceptance | This is concerned with the acceptance of the software and involves acceptance testing to ensure that the tested software is fit for purpose |
| Rollout | This is concerned with the deployment of the software and support/maintenance activities |

[a]There may be additional requirements for public procurement to ensure fairness in the process

## 8.2 Planning and Requirements

The decision on whether the project team should (or has the competence to) develop and test the developed software, or whether there is a need to outsource the work to a specialized third-party supplier is made early in the project. In some situations, the complete project is outsourced to a supplier, and the outsourcing provides the full solution to the project's requirements, or there may need to integrate the solution with other software. In other situations, there may be a partial outsourcing of part of the project, and in this chapter, we consider the activities involved in outsourcing the testing part of the project. The following tasks are involved in test outsourcing:

– The requirements for the outsourcing are defined
– The solution may be to outsource all or part of the testing
– An evaluation team is formed
– Evaluation criteria is prepared
– The approach to procurement is defined
– The initial risks are identified and managed
– The procurement plan is prepared
– The procurement schedule is prepared.

Once the decision has been made to outsource, an evaluation team is formed to identify potential suppliers, and evaluation criteria is defined to enable each supplier's solution to be objectively rated.

A plan will be prepared by the project manager (with assistance from the test manager where outsourcing of testing is involved) detailing the approach to the procurement, defining how the evaluation will be conducted, defining the members of the evaluation team and their roles and responsibilities, and preparing a schedule of the procurement activities to be carried out.

## 8.3 Identifying Suppliers

The list of potential suppliers may be determined in various ways such as previous working relationships, research, recommendations

– Previous working relationship with suppliers
– Research via the Internet/Gartner
– Recommendations from colleagues or another company
– Advertisements/other.

The fact that a supplier has worked previously with the company is valuable, as it provides useful information on the capability of the supplier, and whether it would be a good fit for the work to be done. Companies will often maintain a list of *preferred suppliers*, and these are the suppliers that have worked previously with the company, and whose capability is known. The risks associated with a supplier on the preferred supplier list are generally lower than those of an unknown supplier. If the experience of working with the supplier is poor, then the supplier may be removed from the preferred supplier list.

There may be additional requirements for public procurement to ensure fairness in the procurement process, and often public contracts need to be more widely advertised. For example, public contracts for services or supplies in the European Union that are over a certain threshold value need to be published in the Official Journal of the European Union (OJEU) to allow all interested parties the opportunity to make a proposal to provide the product or service.

The list of candidate suppliers may potentially be quite large, and so shortlisting may be employed to reduce the list to a more manageable size of candidate suppliers.

## 8.4 Prepare and Issue RFP

Once the candidate suppliers have been identified, the procurement team will contact the potential suppliers advising them of the work that is available and requesting them to propose their solution. The request for proposal (RFP) is prepared and issued to the potential suppliers, and the suppliers are required to complete a proposal detailing their planned solution, as well as the associated costs, by the closing date. The proposal will need to detail the specifics of the supplier's solution, as well as how the supplier plans to perform the required testing.

The RFP details the requirements for the testing and must contain sufficient information to allow the candidate supplier to provide a complete and accurate response. The completed proposal will include technical and financial information so that a rigorous evaluation of each received proposal may be carried out. The RFP may include the criteria defined to evaluate the supplier, and often weightings are employed to reflect the importance of individual criteria. The evaluation criteria may include several categories such as

- Functional (related to business requirements)
- Technology (related to the technologies/test tools)
- Supplier capability and maturity
- Approach
- Overall cost.

Once the proposals have been received, further shortlisting may take place to limit the formal evaluation to around 3–5 suppliers.

## 8.5 Evaluate Proposals and Select Supplier

The evaluation team will evaluate all received proposals using an evaluation spreadsheet (or similar mechanism), and the results of the evaluation yield a shortlist of around three suppliers. The shortlisted suppliers are then invited to make a presentation to the evaluation team, which allows the team to question each supplier in detail to gain a better understanding of the solution that they are offering. This helps in identifying any risks with the supplier and their proposed solution.

Following the presentations and Q&A sessions, the evaluation team will follow up with checks on reference sites for each supplier. The evaluation spread sheet is updated with all the information gained from the presentations, the reference site checks, and the risks associated with the suppliers.

Finally, an evaluation report is prepared to give a summary of the evaluation, and this includes the recommendation of the preferred supplier. The project board then makes a decision to accept the recommendation; select an alternate supplier; or restart the procurement process.

## 8.6 Formal Agreement

The preferred supplier is informed on the outcome of the evaluation and selection, and negotiations on a formal legal agreement commences (Fig. 8.1). The agreement will need to be signed by both parties, and it may (depending on the type of agreement) include

- Legal contract
- Statement of work
- Implementation plan
- Training plan
- User guides and manuals

**Fig. 8.1** Legal agreement

- Customer support to be provided
- Service level agreement
- Escrow agreement
- Warranty period.

The *statement of work* (SOW) is employed in bespoke software development, and it details the work to be carried out, the activities involved, the deliverables to be produced, the personnel involved and their roles and responsibilities.

A *service level agreement* (SLA) is an agreement between the customer and service provider which specifies the service that the customer will receive as well as the response time to customer issues and problems. It will also detail the penalties should the service performance fall below the defined levels.

An *escrow agreement* is an agreement made between two parties where an independent trusted third party acts as an intermediary between both parties. The intermediary receives money from one party and sends it to the other party when contractual obligations are satisfied. Under an escrow agreement, the trusted third party may also hold documents and project deliverables such as source code and test plans.

## 8.7 Managing the Supplier

The activities involved in the management of the supplier are similar to the standard test monitoring and control activities discussed in Chap. 7. The outsourcing of the testing may be to a supplier based in a different physical location (possibly in another country), and so regular communication is essential for the duration of the project. The communication may be actual meetings with the supplier if the supplier is on-site, and other forms of communication such as regular status reports, telephone calls, conference calls, and Skype calls when the supplier is in a different physical location.

The project manager is responsible for managing the supplier and will communicate with the supplier on a regular (often daily) basis. The supplier will send regular status reports detailing the progress made in testing as well as any risks and issues. The activities involved in the management of the supplier include

- Monitoring progress
- Managing schedule, effort, and budget
- Managing risks and issues
- Managing changes to the scope of the project
- Obtaining weekly progress reports from the supplier
- Managing key milestones
- Managing quality

– Reviewing the supplier's work
– Performing audits of the test deliverables
– Monitoring test results
– Acceptance testing.

The project manager will maintain daily contact with the supplier and will monitor progress, milestones, risks, and issues. There are risks associated with the supplier such as the supplier delivering late or delivering poor quality, and all supplier risks need to be managed.

## 8.8 Acceptance Testing

The customer carries out user acceptance testing to ensure that the software provided is fit for purpose. The software may just be a part of the overall system, and it may need to be integrated with other software. The acceptance testing involves

– Preparation of acceptance test cases (this is the acceptance criteria)
– Planning and scheduling of acceptance testing
– Setting up the acceptance test environment
– Execution of test cases (UAT testing) to verify acceptance criteria is satisfied
– Test reporting
– Communication of defects
– Correction of the defects
– Re-testing and acceptance of software.

The project manager will communicate any defects with the software to the project team, and the supplier performs all required testing to verify that the software is correct. Once all acceptance tests have successfully passed, the software is accepted.

## 8.9 Rollout and Customer Support

This activity is concerned with the rollout of the software at the customer site, and the handover to the support and maintenance team. It involves

– Deployment of the software at customer site
– Provision of training to staff
– Handover to the support and maintenance team.

## 8.10 Review Questions

1. What are the main activities in the selection and management of a supplier for test outsourcing?
2. What factors would lead an organization to seek a supplier rather than testing the software solution in-house?
3. What are the benefits of outsourcing?
4. Describe how a supplier should be selected.
5. Describe how a supplier should be managed.
6. What is a service level agreement?
7. Describe the purpose of the statement of work?
8. What is an escrow agreement?

## 8.11 Summary

Supplier selection and management is concerned with the selection and management of a third-party software supplier. Many large projects often involve total or partial outsourcing of the software development or testing, and it is therefore essential to select a supplier that is capable of delivering a high-quality and reliable solution on time and on budget.

This means that the process for the selection of the supplier needs to be rigorous, and that the capability of the supplier is clearly understood, as well as being aware of any risks associated with the supplier. The selection is based on objective criteria, and the evaluation team will rate each supplier against the criteria and recommend their preferred supplier.

Once the selection is finalized, a legal agreement is drawn up (which usually includes the terms and condition of the contract as well as a statement of work). The supplier then commences the defined work and is appropriately managed for the duration of the contract.

The project manager is responsible for managing the supplier, and this involves communicating with the supplier on a daily basis and managing issues and risks. The software is subject to acceptance testing before it is accepted from the supplier.

# Test Metrics and Problem-Solving

# 9

**Key Topics**

Measurement
Goal, question, metric
Problem-solving
Data gathering
Fishbone diagram
Histogram
Pareto chart
Trend graph
Scatter graph
Statistical process control

## 9.1 Introduction

Measurement is an essential part of mathematics and the physical sciences, and it has been successfully applied to the software engineering field. The purpose of a measurement program is to establish and use quantitative measurements to manage the software development projects and software quality in an organization; to assist the organization in understanding its current software engineering capability; and to provide an objective indication that software process improvements have been successful.

Measurements provide visibility into the various functional areas in the organization, and the quantitative data allow trends to be seen over time. The analysis of the measurements allows action plans to be produced for continuous improvement. Measurements may be employed to track the quality, timeliness, cost, schedule, and



G. O'Regan, *Concise Guide to Software Testing*,
Undergraduate Topics in Computer Science,
https://doi.org/10.1007/978-3-030-28494-7_9

effort of software projects. The terms "*metric*" and "*measurement*" are used interchangeably in this book. The formal definition of measurement given by Fenton (1995) is:

> Measurement is the process by which numbers or symbols are assigned to attributes or entities in the real world in such a way as to describe them according to clearly defined rules.

Measurement plays a key role in the physical sciences and everyday life: for example, calculating the distance to the planets and stars; determining the mass of objects; computing the speed of mechanical vehicles; calculating the electric current flowing through a wire; computing the rate of inflation; estimating the unemployment rate; and so on. Measurement provides a more precise understanding of the entity under study.

Often several measurements are used to provide a detailed understanding of the entity under study. For example, the cockpit of an airplane contains measurements of altitude, speed, temperature, fuel, latitude, longitude, and various devices essential to modern navigation and aviation, and clearly an airline offering to fly passengers using just the altitude measurement alone would not be taken seriously.

Metrics play a key role in problem-solving, and various problem-solving techniques will be discussed later in the chapter. Measurement data provides a quantitative account of the extent of the problem. For example, the elapsed time between the down time and the subsequent up time is a measure of how serious a telecommunication outage is, and it is essential to minimize outages and their impact should one occur. Measurements may be used as part of the analysis on the root cause of a particular problem, e.g. of a telecommunications outage and to verify that the actions taken to correct the problem have been effective.

Metrics may be used to provide an internal view of the quality of the software product, but care is needed before deducing the behaviour that a product will exhibit externally from the various internal measurements of the product. A *leading measure* is a software measure that usually precedes the attribute that is under examination; for example, the arrival rate of software problems is a leading indicator of the maintenance effort. Leading measures provide an indication of the likely behaviour of the product in the field and need to be examined closely. A *lagging indicator* is a software measure that is likely to follow the attribute being studied; for example, escaped customer defects are an indicator of the quality and reliability of the software. It is important to learn from lagging indicators even if the data can have little impact on the current project.

## 9.2 The Goal, Question, Metric Paradigm

Many software metrics programs have failed due to poorly defined, or non-existent goals and objectives, with the metrics defined unrelated to the achievement of the business goals. The *Goal, Question, Metric* (GQM) paradigm was developed by

Victor Basili and others of the University of Maryland in the late 1980s (Basili and Rombach 1988). It is a rigorous goal-oriented approach to measurement, in which goals, questions, and measurements are closely integrated.

The business goals are first defined, and then questions that relate to the achievement of the goal are identified. Next, for each question a metric that gives an objective answer is defined. The statement of the business goal is precise, and it is related to individuals or groups. The GQM approach proceeds as follows:

- Set goals specific to needs in terms of purpose, perspective, and environment
- Refine the goals into quantifiable questions
- Determine the metrics and data to be collected (and the means for collecting them) to answer the questions.

GQM has been applied to several domains, and so we consider an example from the software field that aims to determine the effectiveness of a new programming language *L*. There are several valid questions that may be asked at this stage, including who are the programmers that use L? What is their level of experience? What is the quality of software code produced? What is the code productivity of the language? This leads to the quality and productivity metrics as detailed in Fig. 9.1.

*Goal*

The focus on improvements should be closely related to the business goals, and the first step is to identify the key goals that are essential for business success (or to the success of an improvement program). It does not make sense to direct improvement activities to areas that do not require improvement, or for which there is no business need to improve, or from which there will be a minimal return to the organization.

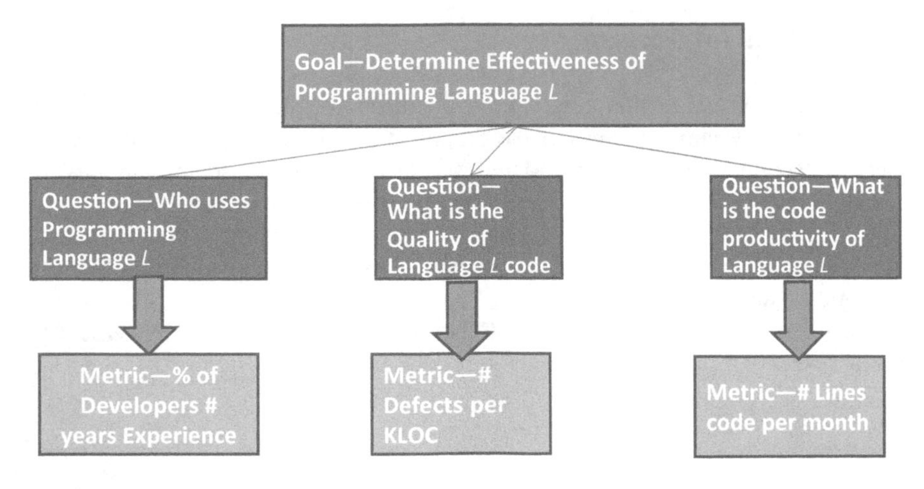


**Fig. 9.1** GQM example

*Question*
These are the key questions that determine the extent to which the goal is being satisfied, and for each business goal the set of pertinent questions need to be identified. Each question is analysed to determine the best approach to obtain an objective answer, and to define the metrics that are needed, and the data that needs to be gathered to answer the question objectively.

*Metrics*
These are measurements that give a quantitative answer to the particular question, and they provide an objective picture of the extent to which the goal is currently satisfied. Measurement improves the understanding of a specific process or product, and the GQM approach leads to measurements that are closely related to the goal, *rather than measurement for the sake of measurement.*

GQM helps to ensure that the defined measurements will be relevant and used by the organizations to understand its current performance and to improve and satisfy its business goals more effectively. It is a rigorous approach to software measurement, and the measures may be from various viewpoints, e.g. manager viewpoint, project team viewpoint, etc. The idea is always first to identify the goals, and once the goals have been decided common-sense questions and measurement are employed.

There are two key approaches to software process improvement: *top-down* or *bottom-up* improvement. Top-down approaches are based on process improvement models and appraisals: e.g. models such as the CMMI, ISO 15504, and ISO 9000, whereas GQM is a bottom-up approach to software process improvement and is focused on improvements related to specific goals. The top-down and bottom-up approaches are often combined in practice.

## 9.3 Metrics for Testing

The objective of this section is to present a collection of test and other metrics to provide visibility into key areas of the organization and to show how metrics are used to facilitate improvement. Many organizations have monthly quality or operation reviews in which the presentation of metrics plays an important part, and where improvement trends in the metrics may be seen over time. The main output from a management review is a series of improvement actions, which should result in tangible improvements.

We present sample metrics for customer satisfaction, project management of testing, execution of testing, customer care, and the cost of quality.

### 9.3.1 Customer Satisfaction Metrics

Figure 9.2 shows the customer survey arrival rate per customer per month, and it shows that there is a customer satisfaction process in place in the organization, that the customers are surveyed, and the extent to which they are surveyed.

It does not provide any information as to whether the customers are satisfied, whether any follow-up activity from the survey is required, or whether the frequency of surveys is sufficient (or excessive) for the organization. Figure 9.3 gives the customer satisfaction measurements for a particular customer, and it contains several categories such as quality, timeliness in meeting the committed dates, ability to deliver the agreed content, the ease of use of the software, the expertise of the staff and the value for money. The numerical interpretation is:

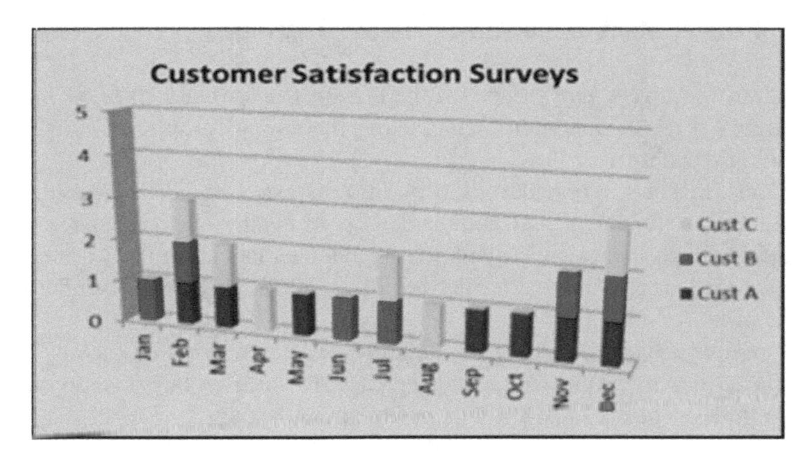


**Fig. 9.2** Customer survey arrivals

**Fig. 9.3** Customer satisfaction measurement

| | |
|---|---|
| 8–10 | Exceeds expectations |
| 7 | Meets expectations |
| 5–6 | Fair |
| 0–4 | Below expectations |

Another words, a score of 8 for testing indicates that the customers consider the software to be well tested, and a score of 9 for value for money indicates that the customers considers the solution to be excellent value. It is essential that the customer feedback is analysed (with follow-up meetings held with the customer where appropriate). There may be a need to take corrective action to deal with customer issues, and this may involve producing an action plan, communicating it to the customer, and executing the plan.

### 9.3.2 Project Management Metrics for Testing

The metrics for project management of the testing is to provide visibility into the effectiveness of the test manager in completing the testing on time, on budget, and with the right quality.

The timeliness metric provides visibility into the extent to which the testing has been delivered on time (Fig. 9.4), and the number of months over or under schedule per project in the organization is shown. The schedule timeliness metric is a lagging measure, as it indicates that the testing has been delivered within schedule or not after the event.

The on-time delivery of testing during a project requires careful tracking of the various activities in testing, and corrective actions need to be taken to address slippage in development or delays that occur during testing.

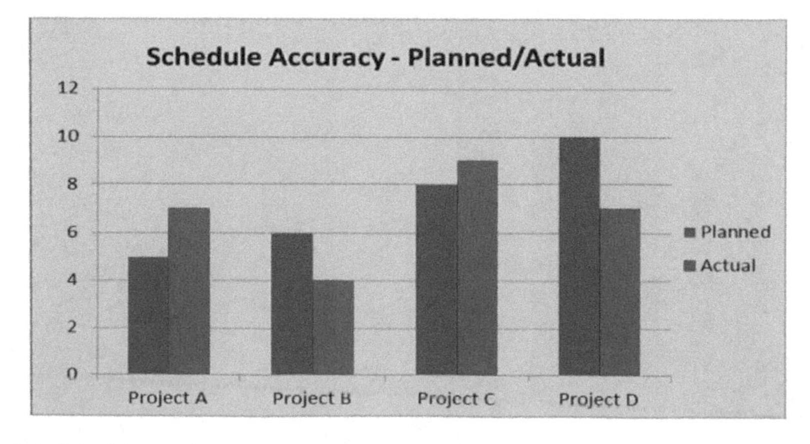


**Fig. 9.4** Schedule timeliness metric for testing

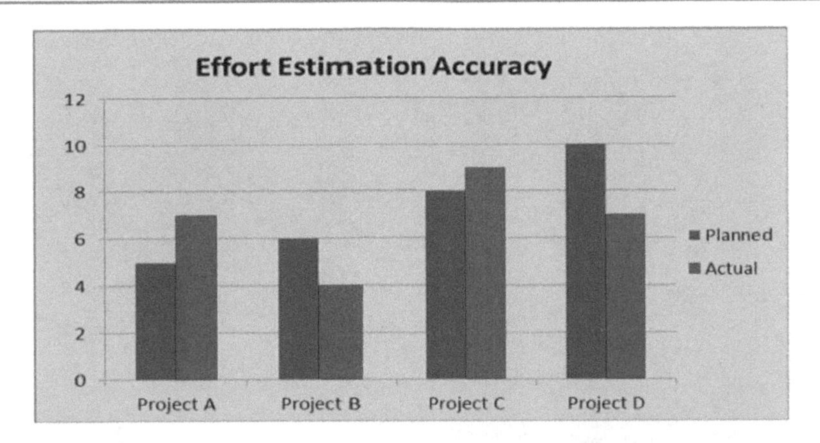


**Fig. 9.5** Effort timeliness metric for testing

The next metric provides visibility into the effort estimation accuracy of the testing (Fig. 9.5). Effort estimation is a key component in calculating the cost of testing and in preparing the schedule, and accurate estimation is a challenge.

The effort estimation chart is similar to the schedule estimation chart, except that the schedule metric is referring to time as recorded in elapsed calendar months, whereas the effort estimation chart refers to the planned number of person months required to carry out the work, and the actual number of person months that it actually took. Projects need an effective estimation methodology to enable them to be successful in project management, and the project (or test) manager will use metrics to determine how accurate the estimation has actually been.

### 9.3.3 Test Execution Metrics

These metrics give visibility into the testing, and Fig. 9.6 gives an indication of the quality of the software produced, and the quality of the definition of the initial requirements. It shows the total number of defects and the total number of change requests raised during the project, as well as details on their severities. The presence of a large number of change requests suggests that the initial definition of the requirement was incomplete, and that there is room for improvement in the requirements process.

Figure 9.7 gives the status of open defects and change requests with the project, which gives an indication of the current quality of the project, and the effort required to achieve the desired quality in the software. This chart is not used in isolation, as the test manager will need to know the arrival rate of problems (Figs. 9.9 and 9.12) to determine the stability of the software product.

The organization may decide to release a software product with open problems provided that the associated risks with the known problems can be managed. It is essential to perform a risk assessment of all known problems and to document the open problems (with their workarounds) in the release notes.

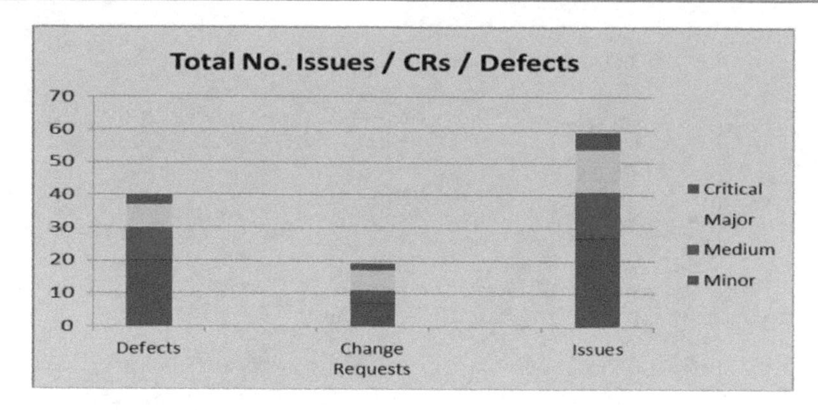


**Fig. 9.6** Total number of issues in project

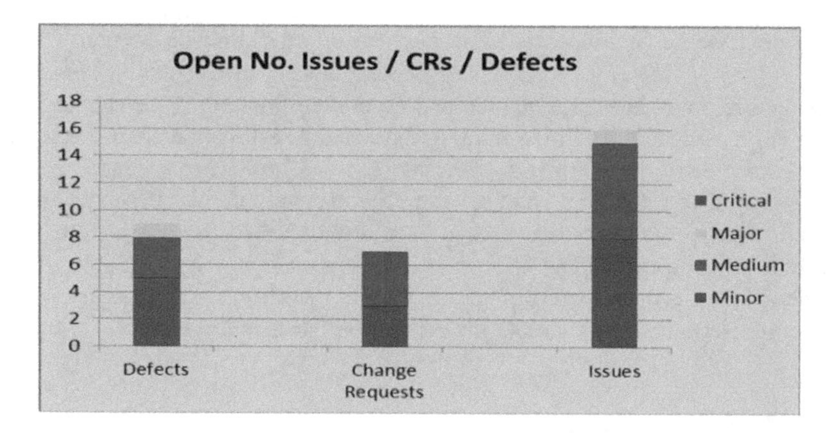


**Fig. 9.7** Open issues in project

The test manager will need to know the age of the open problems to determine the effectiveness of the project team in resolving problems in a timely manner. Figure 9.8 presents a metric to present the age of the open defects, and it highlights the fact that there is one major problem that has been open for over one year. The project manager needs to prevent this situation from arising, as critical and major problems should be swiftly resolved.

The problem arrival rate enables the test manager to judge the stability of the software, and this (along with other metrics) helps in judging whether the software is fit for purpose and ready for release to potential customers. Figure 9.9 presents a sample problem arrival chart, which indicates positive trends with the arrival rate falling to very low levels.

The test manager will need to do analysis to determine if there are other causes that could contribute to the fall in the arrival rate; for example, it may be the case that testing was completed in September, which would mean, in effect, that no testing has been performed since then, with an inevitable fall in the number of

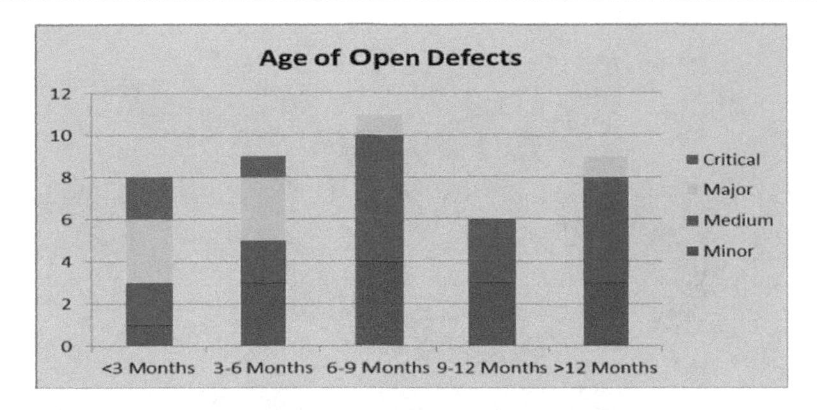


**Fig. 9.8** Age of open defects in project

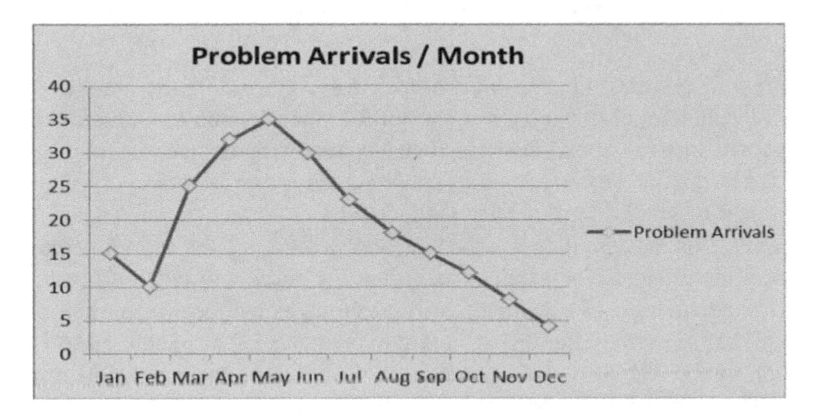


**Fig. 9.9** Problem arrivals per month

problems reported. The important point is not to jump to a conclusion based on a particular chart, as the circumstances behind the chart should be fully known and taken into account in order to draw valid conclusions.

Figure 9.10 measures the effectiveness of the project in identifying defects in the development phase, and the effectiveness of the test groups in detecting defects that are present in the software. The development portion typically includes defects reported on inspection forms and in unit testing.

The chart indicates that the project had a phase containment effectiveness of approximately 54%. That is, the developers identified 54% of the defects, the system testing phase identified approximately 23% of the defects, acceptance testing identified approximately 14% of the defects, and the customer identified approximately 9% of the defects.

The objective is that the number of defects reported at acceptance test and after the product is officially released to customer should be minimal (preferably zero defects post release of the software).

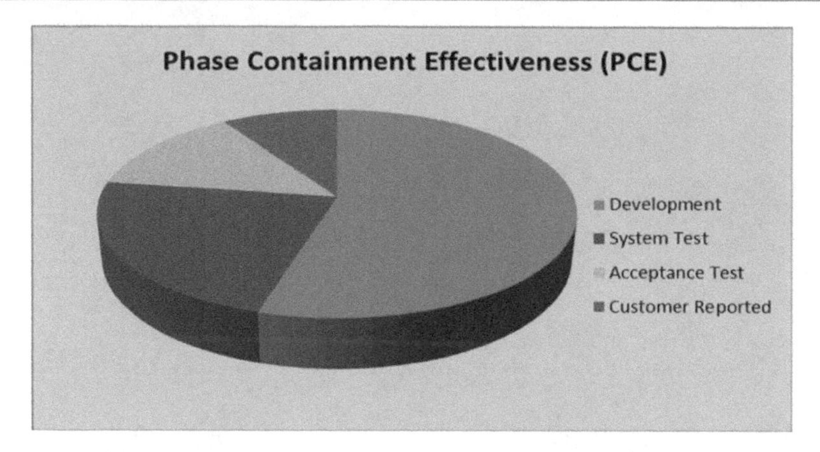


Fig. 9.10 Phase containment effectiveness

Figure 9.11 presents the test status of the project, including the number of tests planned, the number of test cases run, the number that have passed, and the number of failed and blocked tests. The test status is reported regularly to management during the testing, and extra resources are provided where necessary to ensure that the customer receives a high-quality product with all defects corrected.

Figure 9.12 is the cumulative arrival rate curve, and it gives an indication of the stability of the product. The expectation is that the curve will level off towards the end of testing, as most of the defects will previously have been identified.

Figure 9.13 describes the arrival and closure rates of problems and gives an indication of the stability of the project as well as its effectiveness in resolving defects. The arrival rate of problems should be very low towards the end of the project.

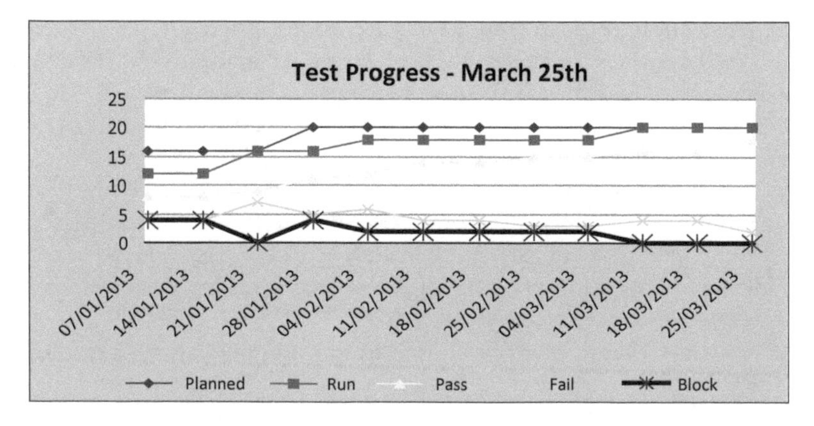


Fig. 9.11 Test status

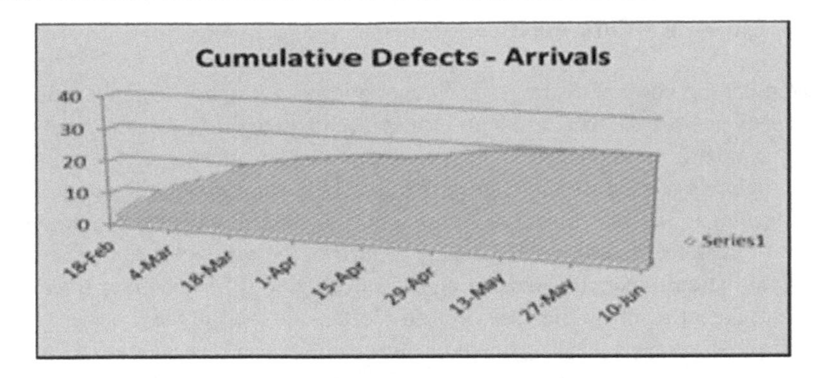


**Fig. 9.12** Cumulative defects

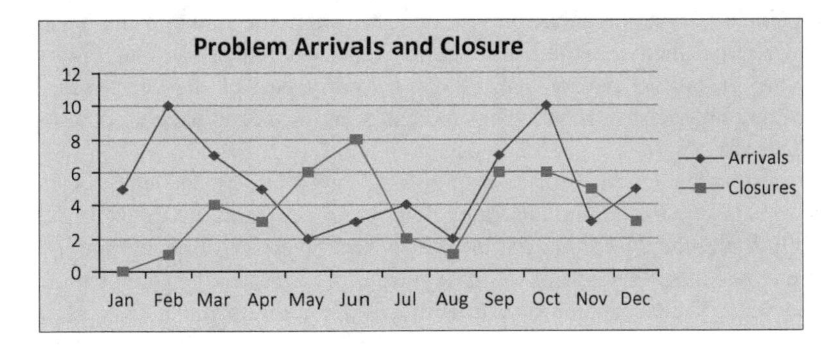


**Fig. 9.13** Problem arrival and closure

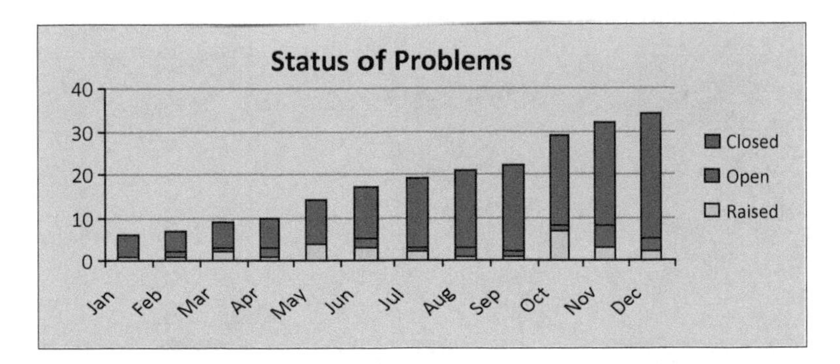


**Fig. 9.14** Status of problem

Figure 9.14 gives an indication of the number of raised, open and closed problems during the project. It does not give an indication of how serious the problems are.

### 9.3.4 Customer Care Metrics

The goals of the customer care group in an organization are to respond efficiently and effectively to customer problems, to ensure that their customers receive the highest standards of service from the company, and to ensure that its products function reliably at the customer's site. The customer care group will need to know how effective it is in resolving customer queries, and it will need to know the number of customer queries raised during a period, the availability of its software systems at the customer site, and the age of open queries. A customer query may result in a defect report in the case of a problem with the software.

Figure 9.15 presents the arrival and closure rate of customer queries (it could be developed further to include a severity attribute for the query). Quantitative goals are generally set for the resolution of queries (especially where there is a service level agreement in place). A chart for the age of open queries (similar to Fig. 9.8) is often maintained. The organization will need to know the status of the backlog of open queries per month, and a simple trend graph would provide this. Figure 9.15 shows that the arrival rate of queries: in the early part of the year exceeds the closure rate of queries per month. This indicates an increasing backlog that needs to be addressed.

The customer care department responds to any outages and ensures that the outage time is kept to a minimum. Many companies set ambitious goals for network availability: e.g. the "*five nines initiative*" has the objective of developing systems which are available 99.999% of the time, i.e. approximately five minutes of down time per year. The calculation of availability is from the formula:

$$\text{Availability} = \frac{\text{MTBF}}{\text{MTBF} + \text{MTTR}}$$

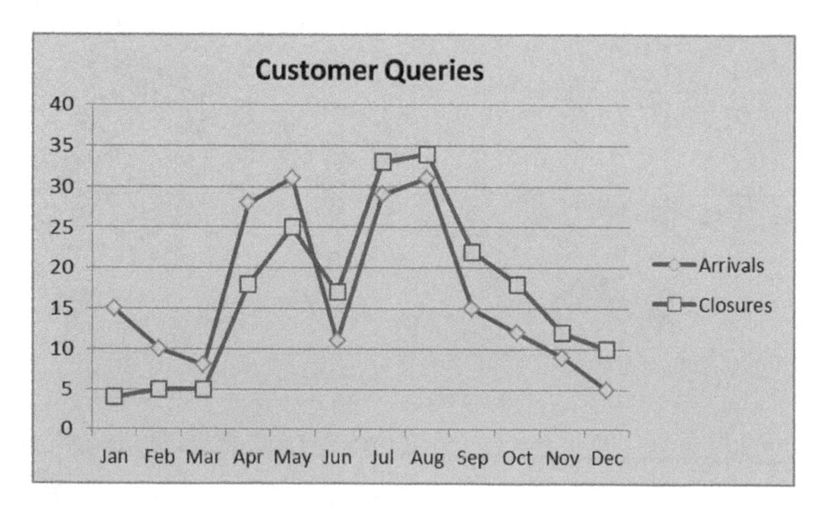


**Fig. 9.15** Customer queries (arrivals/closures)

where the mean time between failure (MTBF) is the average length of time between outages.

$$\text{MTBF} = \frac{\text{Sample Intervel Time}}{\#\text{Outages}}$$

The formula for MTBF above is for a single system only, and the formula is adjusted when there are multiple systems.

$$\text{MTBF} = \frac{\text{Sample Intervel Time}}{\#\text{Outages}} * \#\text{Systems}$$

The mean time to repair (MTTR) is the average length of time that it takes to correct the outage, i.e. the average duration of the outages that have occurred, and it is calculated from the following formula:

$$\text{MTTR} = \frac{\text{Total Outage Time}}{\#\text{Outages}}$$

Figure 9.16 presents outage information on the customers impacted by an outage during the particular month and the extent of the impact on the customer.

The customer care department will ensure that a post-mortem of the outage is performed to ensure that lessons are learned to prevent a reoccurrence. The causal analysis identifies the root causes of the outage, and corrective actions are implemented to prevent a reoccurrence. Metrics to record the amount of system availability and outage time per month will be maintained by the customer care group, and Fig. 9.17 provides visibility on the availability of the system.

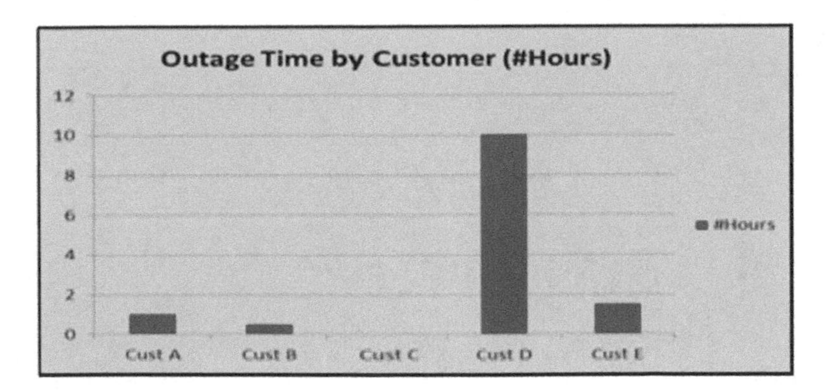


**Fig. 9.16** Outage time per customer

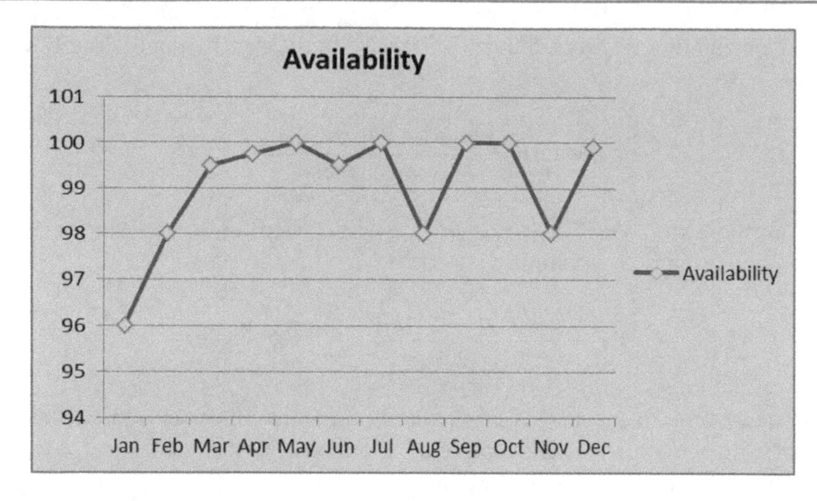


**Fig. 9.17** Availability of system per month

### 9.3.5 Miscellaneous Metrics

Metrics may be applied to many other areas in the organization. This section includes metrics on the CMMI maturity of an organization (where an organization is implementing the CMMI) and the cost of poor quality. Figure 9.18 gives the internal CMMI maturity of the organization and indicates its readiness for a formal CMMI assessment. A numeric score of 1–10 is used to rate each process area, and a score of 7 or above indicates that the process area is satisfied.

Crosby argued that the most meaningful measurement of quality is the cost of poor quality (Crosby 1979), and that the emphasis on the improvement activities in the organization should therefore be to reduce the *cost of poor quality* (COPQ).

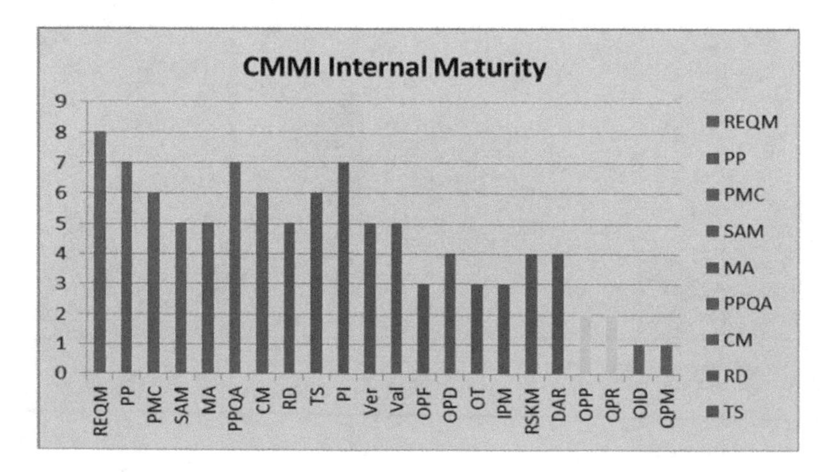


**Fig. 9.18** CMMI maturity in current year

The cost of quality includes the cost of external and internal failure, the cost of providing an infrastructure to prevent the occurrence of problems, and the cost of the infrastructure to verify the correctness of the product (see Chap.1).

The cost of quality was divided into four subcategories by Feigenbaum in the 1950s:

- Cost external
- Cost internal
- Cost prevention
- Cost appraisal.

The cost of quality graph (Fig. 1.9) will initially show high external and internal costs and low prevention costs, and the total cost of quality will be high. However, as an effective quality system is put in place and becomes operational, there will be a noticeable decrease in the external and internal cost of quality, and a gradual increase in the cost of prevention and appraisal.

The total cost of quality will substantially decrease, as the cost of provision of the quality system is substantially below the cost of internal and external failure. The COPQ curve indicates where the organization is in relation to the cost of poor quality, and the organization will need to implement improvements to put an effective quality management system in place to minimize the cost of poor quality.

## 9.4 Implementing a Metrics Program

The metrics that we have discussed in this chapter may be adapted and tailored to meet the needs of organizations. The metrics are only as good as the underlying data, and good data gathering is essential. Table 9.1 describes typical steps in the implementation of a metrics program:

The business goals are the starting point in the implementation of a metrics program, as there is no sense in measurement for the sake of measurement. The next step is to identify the relevant questions to determine the extent to which the business goal is being achieved and to define metrics that provide an objective answer to the questions.

The organization defines its business goals, and each department develops specific goals to support the business goals. Next, the data to be gathered and the methods by which the data may be recorded are determined. A small organization may record the data manually, but often automated or semi-automated tools will be employed in larger organisations. It is essential that the data collection and extraction are efficient, as otherwise the metrics program may fail.

The roles and responsibilities of staff with respect to the implementation and day-to-day operation of the metrics program need to be defined, and staff trained to perform their roles effectively. Finally, a regular management review is needed,

**Table 9.1** Implementing metrics

| Implementing metrics in organization |
|---|
| Define the business goals |
| Determine questions related to achievement of goals |
| Define the metrics |
| Determine data that needs to be gathered |
| Identify tools to (semi-) automate metrics |
| Identify and provide needed resources |
| Gather data and prepare metrics |
| Communicate the metrics and review monthly |
| Provide training |

where the metrics and trends are presented, and actions identified and carried out to ensure that the business goals are achieved.

### 9.4.1 Data Gathering for Metrics

Metrics are only as good as the underlying data, and so data gathering is a key activity in the program. The data will be closely related to the questions and used to give an objective answer to the questions. The business goals are usually expressed quantitatively, and Table 9.2 presents an example of how the questions related to a particular goal are identified.

Table 9.3 is designed to determine the effectiveness of the software development process and to enable the above questions to be answered. It includes a column for

**Table 9.2** Goals and questions

| Goal | Reduce escaped defects from each lifecycle phases by 10% |
|---|---|
| Questions | How many defects are identified within each lifecycle phase?<br>How many defects are identified after each lifecycle phase is exited?<br>What percentage of defects escaped from each lifecycle phase? |

**Table 9.3** Phase containment effectiveness

| Phase of origin | | | | | | | | |
|---|---|---|---|---|---|---|---|---|
| Phase | Inspect defects | Reqs | Design | Code | Accept test | In-phase defects | Other defects | % PCE |
| Reqs | 4 | | 1 | 1 | | 4 | 6 | 40% |
| Design | 3 | | | | | 3 | 4 | 42% |
| Code | 20 | | | | | 20 | 15 | 57% |
| Unit test | | 2 | 2 | 10 | | | | |
| System test | | 2 | 2 | 5 | | | | |
| Accept test | | | | | | | | |

inspection data that records the number of *defects* recorded at the various inspections. The *defects* include the phase where the defect originated; for example, a defect identified in the coding phase may have originated in the requirements or design phase. This data is typically maintained in a spreadsheet, e.g. Excel (or a dedicated tool), and it needs to be kept up to date. It enables the phase containment effectiveness (PCE) to be calculated for the various phases.

We distinguish between a defect that is detected *in-phase* versus a defect that is detected *out-of-phase*. An in-phase defect is a problem that is detected in the phase in which it is created (e.g. usually by a software inspection). An out-of-phase defect is detected in a later phase (e.g. a defect with the requirements may be discovered in the design or coding phase: i.e. a later phase from the phrase in which it was created).

The effectiveness of the requirements phase in Table 9.3 is judged by its success in identifying defects as early as possible, as the cost of correction of a requirements defect increases the later in the cycle that it is identified. The requirements PCE is calculated to be 40%, i.e. the total number of defects identified in the requirements phase divided by the total number of requirements defects identified. There were four defects identified at the inspection of the requirements, and six defects were identified outside of the requirements phase: one in the design phase, one in the coding phase, two in the unit testing phase, and two at the system-testing phase: i.e. 4/10 = 40%. Similarly, the code PCE is calculated to be 57%.

The overall PCE for the project is calculated to be the total number of defects detected in phase in the project divided by the total number of defects, i.e. 27/52 = 52%. Table 9.3 is a summary of the collected data and its construction consists of:

- Maintain inspection data of requirements, design, and code inspections
- Identify defects in each phase and determine their phase of origin
- Record the number of defects in each phase per phase of origin.

Software inspections need to record the problems identified and the phase of origin, and staff need to be appropriately trained to do this consistently. The example above gives a flavour of data gathering, and in practice the organization will need to collect various data for the metrics to give an objective answer on the extent to which the particular goal is being satisfied.

## 9.5 Problem-Solving Techniques

Problem-solving is a key part of quality improvement, and a *quality circle* (or problem-solving team) is a group of employees who do similar work and volunteer to come together on company time to identify and analyse work-related problems.

There are several tools to support problem-solving including *process mapping*, *trend charts*, *bar charts*, *scatter diagrams*, *fishbone diagrams*, *histograms*, *control charts*, and *Pareto charts* (Brassard and Ritter 1994). These provide visibility into the extent of the problem and a problem-solving team is:

- A group of employees who do similar work
- Voluntarily meet regularly on company time
- Circle leader acts as a facilitator
- Identify and analyse work-related problems
- Recommend solutions to management
- Implement solution where possible.

The facilitator of the quality circle coordinates the activities, ensures that the team members receive sufficient training, and obtains specialist help where required. The leader has the following responsibilities:

- Focal point of quality circle activities
- Train team members
- Coordinate activities of the circle group
- Assist in inter-circle investigations
- Obtain specialist help when required.

The circle leaders receive training in problem-solving techniques and are responsible for training the team members. The circle leader needs to keep the meeting focused and requires skills in team building. The steps in problem-solving include

- Select the problem
- State and restate the problem
- Collect the facts
- Brainstorm
- Choose course of action
- Present to management
- Measurement of success.

The benefits of a successful problem-solving culture in the organization include

- Savings of time and money
- Increased productivity
- Reduced defects
- Fire prevention culture.

Several problem-solving tools are discussed in the following sections.

### 9.5.1 Fishbone Diagram

This well-known problem-solving tool consists of a cause-and-effect diagram that is in the shape of the backbone of a fish. The objective is to identify the various causes of some particular problem, and these causes are then broken down into a number of subcauses. The various causes and subcauses are analysed to determine the root cause of the particular problem, and actions to address the root cause are then defined to prevent a reoccurrence. There are various categories of causes, and these may include people, methods and tools, and training.

The fishbone diagram offers a crisp mechanism to summarize the collective knowledge that a team has about a particular problem, as it focuses on the causes of the problem and facilitates the detailed exploration of the causes (Fig. 9.19). Its construction involves a clear statement of the particular effect, which is placed at the right-hand side of the diagram. The major categories of cause are drawn on the backbone of the fishbone diagram; brainstorming is used to identify causes; and these are then placed in the appropriate category. For each cause identified the various subcauses are identified by asking the question "*Why does this happen*"? This leads to a more detailed understanding of the causes and subcauses of a particular problem.

*Example 9.1* An organization wishes to determine the causes of a high number of customer reported defects. There are various categories that may be employed such as people, training, methods, tools, and environment. In practice, the fishbone diagram in Fig. 9.19 would be more detailed than that presented, as subcauses would also be identified by a detailed examination of the identified causes. The root cause(s) are determined from detailed analysis.

This example suggests that the organization has significant works to do in several areas (e.g. test tools, training, and morale) and that major improvements are required. These may include the implementation of a software development process

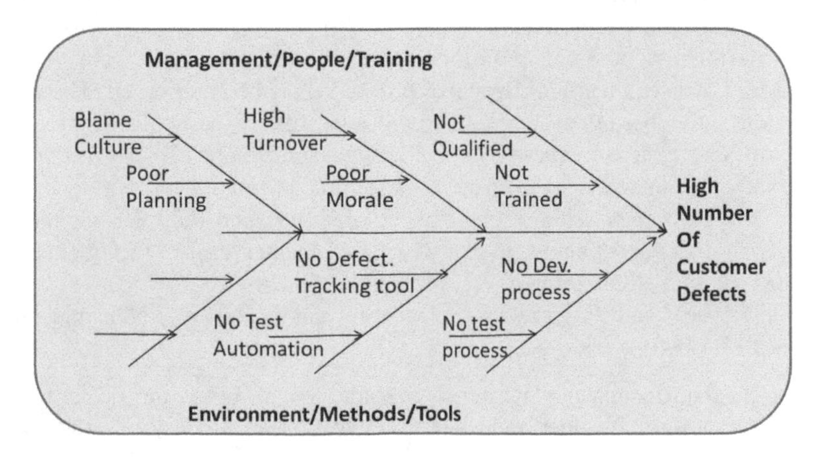


**Fig. 9.19** Fishbone cause-and-effect diagram of high number of defects

and a software test process; the provision of training to enable staff to do their jobs more effectively; and the implementation of better management practices to motivate staff and to provide a supportive environment for software development.

The causes identified may be symptoms rather than actual root causes: for example, high staff turnover may be the result of poor morale and a "blame culture", rather than a cause in itself of poor quality software. The fishbone diagram gives a better understanding of the possible causes of the high number of customer defects. A small subset of these causes is then identified as the root cause(s) of the problem following further discussion and analysis.

The root causes are then addressed by appropriate corrective actions (e.g. an appropriate software development process and test process are defined and training provided on the new processes). The management attitude and organization culture will need to be corrected to enable a supportive software development environment to be put in place.

### 9.5.2 Histograms

A histogram is a way of representing data in bar chart format, and it shows the relative frequency of various data values or ranges of data values. It is usually employed when there are a large number of data values, and it gives a crisp picture of the spread of the data values, and the centring and variance from the mean.

The histogram has an associated shape; e.g. it may be a *normal distribution*, a *bimodal*, or *multi-modal distribution*, or be positively or negatively skewed. The variation and centring refer to the spread of data, and the relation of the centre of the histogram to the customer requirements. The spread of the data is important as it indicates whether the process is too variable, or whether it is performing within the requirements. The histogram is termed process centred if its centre coincides with the customer requirements; otherwise the process is too high or too low. A histogram enables predictions of future performance to be made, assuming that the future will resemble the past.

The construction of a histogram first requires that a frequency table be constructed, and this requires that the range of data values be determined. The data is divided into a number of data buckets, where a bucket is a particular range of data values, and the relative frequency of each bucket is displayed in bar format. The number of class intervals or buckets is determined, and the class intervals are defined. The class intervals are mutually disjoint and span the range of the data values. Each data value belongs to exactly one class interval, and the frequency of each class interval is determined.

The histogram is a well-known statistical tool, and its construction is made more concrete with the following example

*Example 9.2* An organization wishes to characterize the behaviour of the process for the resolution of customer queries in order to achieve its customer satisfaction goal.

*Goal*

Resolve all customer queries within 24 h.

*Question*

How effective is the current customer query resolution process?
What action (if any) is required to achieve this goal?

The data class size chosen for the histogram (Fig. 9.20) is six hours, and the data class sizes is of the same in standard histograms (they may be of unequal size for non-standard histograms). The sample mean is 19 h for this example. The histogram shown is based on query resolution data from 36 samples. The organization goal of customer resolution of all queries within 24 h is not met, and the goal is satisfied in (25/36 = 70% for this particular sample).

Further analysis is needed to determine the reasons why 30% of the goals are outside the target 24-h time period. It may prove to be impossible to meet the goal for all queries, and the organization may need to refine the goal to state that instead all critical and major queries will be resolved within 24 h. Alternately, the solution may be to hire more staff.

### 9.5.3 Pareto Chart

The objective of a Pareto chart is to identify and focus on the resolution of problems that have the greatest impact (as *often 20% of the causes are responsible for 80% of the problems*). The problems are classified into various categories, and the

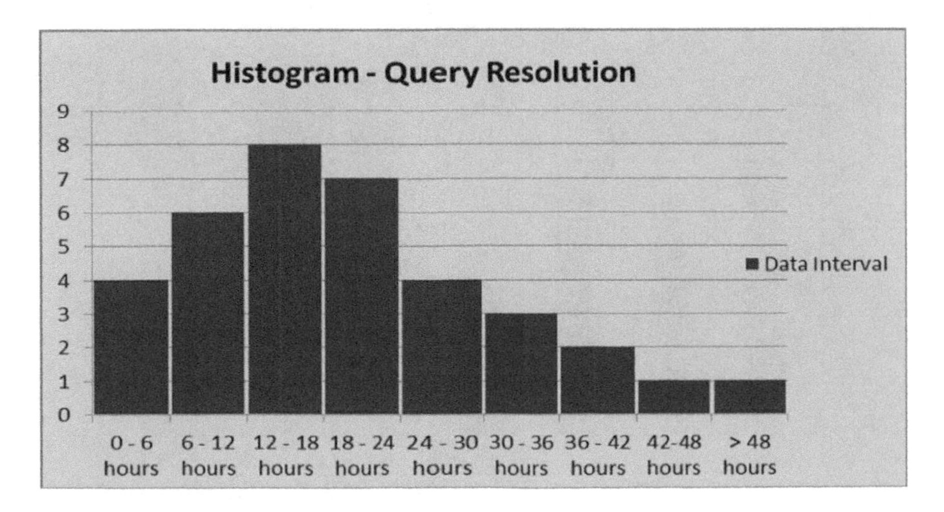


**Fig. 9.20** Histogram

frequency of each category of problem is determined. The Pareto chart is displayed in a descending sequence of frequency, with the most significant cause presented first, and the least significant cause presented last.

The Pareto chart is a key problem-solving tool, and a properly constructed chart will enable the organization to focus on the resolution of the key causes of problems. The effectiveness of the improvements may be judged at a later stage from the analysis of new problems and the creation of a new Pareto chart. The results should show tangible improvements, with less problems arising in the category that was the major source of problems.

The construction of a Pareto chart requires the organization to decide on the problem to be investigated; to identify the causes of the problem via brainstorming; to analyse the historical or real time data; to compute the frequency of each cause; and finally to display the frequency in descending order for each cause category.

*Example 9.3* An organization wishes to understand the various causes of outages and to minimize their occurrence.

The Pareto chart (Fig. 9.21) below includes data from an analysis of outages, where each outage is classified into a particular cause. The six causal categories identified are hardware, software, operator error, power failure, an act of nature, and unknown. The three main causes of outages are hardware, software, and operator error, and analysis is needed to identify appropriate actions to address these. The hardware category may indicate that there are problems with the reliability of the system hardware, and that the existing hardware may need replacement. There may be a need to address availability and reliability concerns with more robust hardware solutions.

The software category may be due to the release of poor quality software, or usability issues with the software, and this requires further investigation.

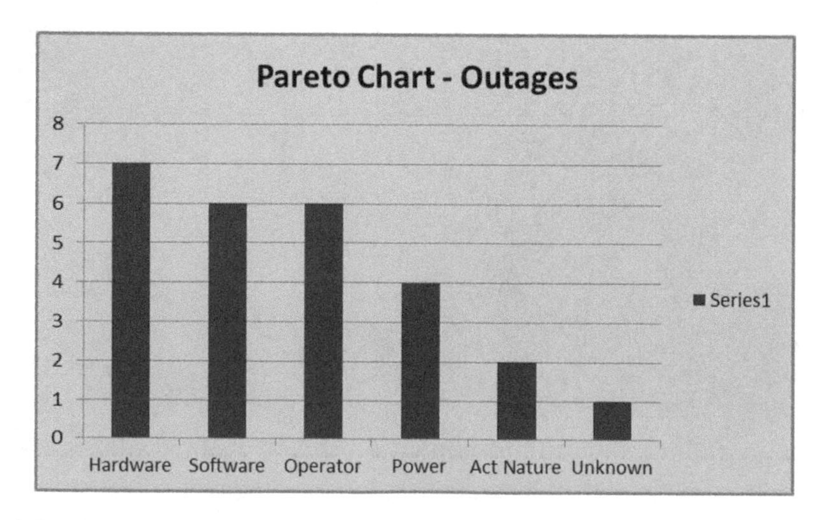


**Fig. 9.21** Pareto chart outages

Finally, operator issues may be due to lack of knowledge or inadequate training of the operators. An improvement plan needs to be prepared and implemented, and its effectiveness will be judged by a reduction in outages, and reductions of problems in the targeted category.

### 9.5.4 Trend Graphs

A trend graph monitors the performance of a variable over time, and it allows trends in performance to be identified, as well as allowing predictions of future trends to be made (assuming that the future resembles the past). Its construction involves deciding on the variable to measure and to gather the data points to plot the data.

*Example 9.4* An organization plans to deploy an enhanced estimation process, and wishes to determine if estimation is actually improving with the new process.

The estimation accuracy determines the extent to which the actual effort differs from the estimated effort. A reading of 25% indicates that the project effort was 25% more than estimated, whereas a reading of −10% indicates that the actual effort was 10% less than estimated.

The trend chart (Fig. 9.22) indicates that initially that estimation accuracy is very poor, but then there is a gradual improvement coinciding with the implementation of the new estimation process.

It is important to analyse the performance trends in the chart. For example, the estimation accuracy for August (17% in the chart) needs to be investigated to determine the reasons why it occurred. It could potentially indicate that a project is using the old estimation process, or that a new project manager received no training on the new process. A trend graph is useful for noting positive or negative trends in performance, with negative trends analysed and actions identified to correct performance.

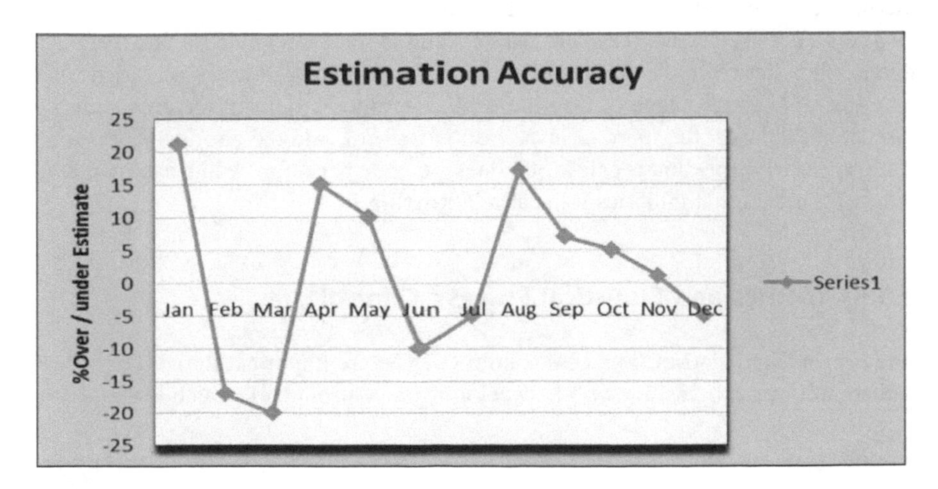


**Fig. 9.22** Trend chart estimation accuracy

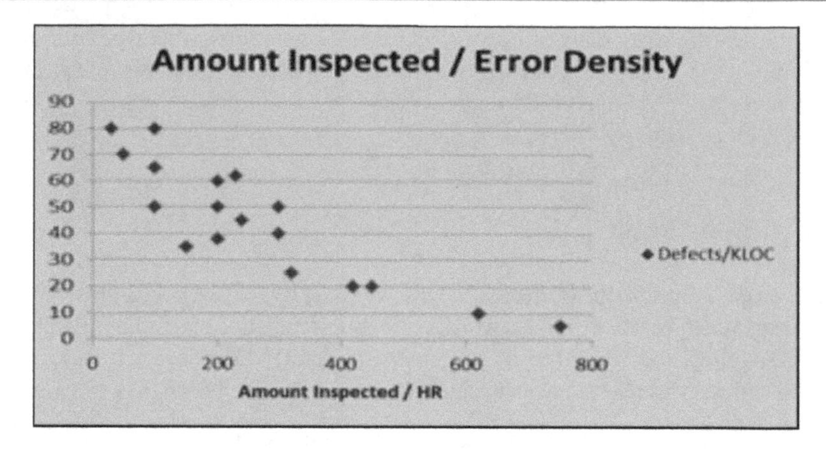


**Fig. 9.23** Scatter graph amount inspected rate/error density

### 9.5.5 Scatter Graphs

The scatter diagram is used to determine whether there is a relationship or correlation between two variables, and if so then to measure the relationship between them. The results may be a positive correlation, negative correlation, or no correlation. Correlation has a precise statistical definition, and it provides a precise mathematical understanding of the extent to which the two variables are related or unrelated.

The scatter graph is often used to determine whether there is a connection between an identified cause and the effect. The construction of a scatter diagram requires the collection of paired samples of data, and the drawing of one variable as the $x$-axis, and the other as the $y$-axis. The data is then plotted and interpreted.

*Example 9.5* An organization wishes to determine if there is a relationship between the inspection rate and the error density of defects identified.

The scatter graph (Fig. 9.23) provides evidence for the hypothesis that there is a relationship between the inspection rates and the error density recorded (per KLOC). The graph suggests that the error density of defects identified during inspections is low if the speed of inspection is too fast, and the error density is high if the speed of inspection is below 300 lines of code per hour. A line can be drawn through the data that indicates a linear relationship.

### 9.5.6 Metrics and Statistical Process Control

The principles of statistical process control (SPC) are important in the monitoring and control of a process. It involves developing a control chart, which is a tool that

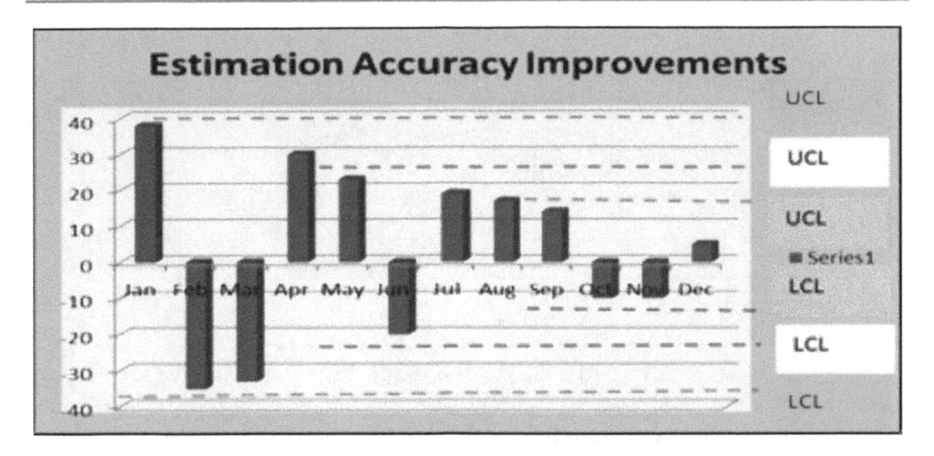


**Fig. 9.24** Estimation accuracy and control charts

may be used to control the process, with upper and lower limits for process performance specified. The process is under control if it is performing within the lower and upper control limits.

Figure 9.24 presents an example on breakthrough in performance of an estimation process and is adapted from Keeni (2000). The initial upper and lower control limits for estimation accuracy are set at ±40%, and the performance of the process is within the defined upper and control limits.

However, the organization revises the upper and lower control limits to ±25%. The organization will need to analyse the slippage data to determine the reasons for the wide variance in the estimation, and part of the solution will be the use of enhanced estimation methods in the organization. In this chart, the organization succeeds in performing within the revised control limit of ±25%, and the limit is revised again to ±15%.

This requires further analysis to determine the causes for slippage and further improvement actions are needed to ensure that the organization performs within the ±15% control limit.

## 9.6 Review Questions

1. Describe the Goal, Question, and Metric model.
2. Describe problem-solving techniques.
3. What is a fishbone diagram?
4. What is a histogram and describe its applications?

5. What is a scatter graph?
6. What is a Pareto chart? Describe its applications.
7. Discuss how a metrics programme may be implemented.
8. What is statistical process control?

## 9.7 Summary

Measurement is an essential part of mathematics and the physical sciences, and it has been successfully applied to the software engineering field. The purpose of a software measurement program is to establish and use quantitative measurements to manage the software development processes, to assist the organization in understanding its current software capability, and to confirm that improvements have been successful.

This chapter included a collection of sample metrics to give visibility into the testing carried out in the organization. It included a presentation of customer satisfaction metrics; test project management metrics; test execution metrics; and customer care metrics.

The Goal, Question, Metric paradigm is a rigorous, goal-oriented approach to measurement in which goals, questions, and measurements are closely integrated. The business goals are first defined, and then questions that relate to the achievement of the goal are identified, and for each question a metric that gives an objective answer to the particular question is defined.

Metrics play a key role in problem-solving, and various problem-solving techniques were discussed. These include histograms, Pareto charts, trend charts, and scatter graphs. The measurement data is used to assist the analysis, to determine the root cause of a particular problem, and to verify that the actions taken to correct the problem have been effective.

Metrics may be employed to track the quality, timeliness, cost, schedule, and effort of software projects. They provide an internal view of the quality of the software product, but care is needed before deducing the behaviour that a product will exhibit externally.

## References

Basili V, Rombach H (1988) The TAME project. Towards improvement-oriented software environments. IEEE Trans Softw Eng 14(6)

Brassard M, Ritter D (1994) The memory jogger. A pocket guide of tools for continuous improvement and effective planning. Goal *I* QPC. Methuen, MA

Crosby P (1979) Quality is free. The art of making quality certain. McGraw Hill, New York
Fenton N (1995) Software metrics: a rigorous approach. Thompson Computer Press
Keeni G, et al (2000) The evolution of quality processes at Tate Consultancy Services. IEEE Softw 17(4)

# Software Testing Tools

# 10

**Key Topics**

Microsoft Project
COCOMO
IBM Rational DOORS
Jira
LDRA Testbed
Integrated development environment
HP Quality Center
Bugzilla
Apache JMeter

## 10.1 Introduction

The goal of this chapter is to give a flavour of a selection of the plethora of tools[1] that are available to support the performance of testing throughout the development lifecycle. Testing plays a key role in verifying that the software system satisfies the requirements and is fit for purpose. There are various categories of test tools ranging from tools that manage software testing to tools that perform specific functions such as automated software inspections or automation of regression tests.

[1]The list of tools discussed in this chapter is intended to give a flavour of what tools are available, and the inclusion of a particular tool is not intended as a recommendation of that tool. Similarly, the omission of a particular tool should not be interpreted as disapproval of that tool.



G. O'Regan, *Concise Guide to Software Testing*,
Undergraduate Topics in Computer Science,
https://doi.org/10.1007/978-3-030-28494-7_10

**Table 10.1** Advantages of test tools

| Advantage | Description |
|---|---|
| Less repetitive work | Test tools reduce the amount of repetitive work in testing. For example, the automation of regression testing (manual regression testing involves entering the same test data several times) or the checking of coding standard with a static analysis tool |
| Consistency | A test tool performs the task exactly as before, whereas there may be slight variations with a human tester |
| Objective | The use of a test tool provides an objective assessment of the testing and helps to ensure that subjective judgments are avoided. For example, test coverage metrics and test execution statistics provide a quantitative measure of how comprehensive the testing has been |
| Visibility | Test tools provide visibility into the testing that has been carried out. These include statistics and graphs of the results of test executions; and reports on defects and performance |
| Audit Trail | Test tools can provide an audit trail of the testing which is useful in the regulated sector |
| Efficiency | Tools can speed up the testing process and help in improving efficiency, consistency, and quality. They support structured testing and save time and effort |

It is a challenge to choose the appropriate tools for a project or organization, and the approach is generally to choose tools to support the process, rather than choosing a process to support the tool.[2] The advantages of test tools include (Table 10.1).

Mature organizations generally employ a structured approach to the introduction of new tools. First, the requirements for a new tool are specified, and the options to satisfy the requirements are considered. These may include developing a tool internally; outsourcing the development of a tool to a third party supplier; or purchasing an off the shelf solution from a vendor. Companies need to ask themselves questions such as

– Is one test tool or several test tools required?
– Should open-source tools be considered (as well as commercial tools)?
– What are the most popular testing tools?
– What lifecycle models are employed?

There are risks that there may be unrealistic expectations in using test tools, including underestimating the time and effort required for the introduction of the new tool. Often it will take a period of time before real benefits will come from the use of the tool. There may also be a lot of time and effort in the maintenance of the test assets related to the tool.

[2]That is, the process normally comes first then the tool rather than the other way around.

**Table 10.2** Tool evaluation table

| | Tool 1 | Tool 2 | … | Tool $k$ |
|---|---|---|---|---|
| Requirement 1 | 8 | 7 | | 9 |
| Requirement 2 | 4 | 6 | | 8 |
| … | | | | |
| … | | | | |
| Requirement $n$ | 3 | 6 | | 8 |
| **Total** | 35 | 38 | … | 45 |

The sample tool evaluation process in Table 10.2 lists all of the requirements vertically that the test tool is to satisfy, and the candidate tools that are to be evaluated and rated against each requirement are listed horizontally. Various rating schemes may be employed, and a simple numeric mechanism is employed in the example. The tool evaluation criteria are used to rate the effectiveness of each candidate tool and indicate the extent to which the tool satisfies the defined requirements. The chosen tool in this example is Tool $k$ as it is the most highly rated of the evaluated tools.

It is normal to consider several candidate tools as part of the selection process, and these may be identified in various ways such as research, word of mouth, and previous working relationships with vendors. Each candidate tool is then evaluated against the criteria to determine the extent to which it satisfies the specified

**Table 10.3** Types of tools for testing

| Type of tool | Description |
|---|---|
| Test management tools | Manage the entire testing process including interfaces to other test tools. It manages the test schedule and test status and enables test results to be logged, and test progress reports to be generated. They also provide traceability between the requirements and test cases and defects |
| Tools for static testing | The software developers use these tools to analyse the software code without executing the code. They enable the developers to understand the structure of the software code and also provide a mechanism to enforce coding standards |
| Test design tools | A test design tool is used to create test input and test cases from the requirements (also design models and GUI). The valid input values are determined from the requirements, and it may be possible to generate the expected results |
| Test execution and logging tools | These include capture/playback tools that often include a scripting language to run the tool, and these tools are especially useful with regression testing. The logging feature allows the test results to be logged |
| Test performance and monitoring tools | Test monitoring tools are used to monitor and track the status of a system over a period of time and to detect problems early. This is used to continuously improve system performance. Test performance tools are used to verify that the performance of the system will meet expected needs |

requirements. An informed decision is then made, and the proposed tool will generally be piloted prior to its deployment. The pilot provides feedback on its suitability, and this will be considered prior to a decision on full deployment. There may need to be customization of the tool to meet the specific needs of the project (or organization) prior to roll out. Finally, the users are trained on the tool, and the tool is rolled out throughout the organization. Support is provided for a period post deployment.

There are various categories of tools to support testing such as test management tools; static testing tools, defect-tracking tools; regression test automation tools; and performance tools (Table 10.3).

## 10.2 Test Management Tools

Test Management tools are used to organize the entire testing process including the planning and scheduling of the various test activities. They allow the team to structure the entire test process with just one test management tool, rather than employing several tools with each tool is performing a specific function for each step of the process. They act as a single application for managing test cases, environments, automated tests, defects, and project reporting. They provide

- Test scheduling
- Management of the test cases
- Traceability
- Results logging and reporting
- Interface to other tools such as requirements management, test execution, defect management, and configuration management tools.

They allow the team to plan the testing activities and report the status of the quality assurance activities. The tools allow the progress of the various tasks to be managed and provide support for test case environments, test case design, and specification, and traceability to the requirements. They provide release management and support test case execution and automated testing, defect logging and tracking. They allow metrics to be generated to give visibility into test progress and quality and provide dashboards that provide a crisp executive summary of the key performance indicators (KPIs) for testing. There are various test management tools available including:

- HP Quality Center
- TestRail
- QTest
- QA Complete.

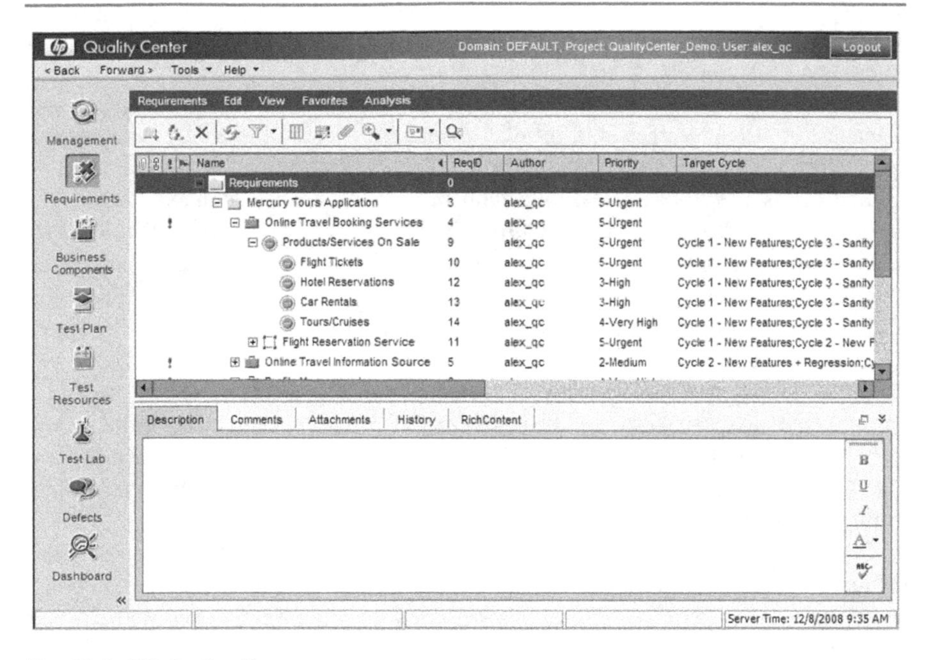


**Fig. 10.1** HP Quality Center

The HP Quality Center™ tool (Fig. 10.1) standardizes and manages the entire test and quality process, and it is a Web-based system test management tool for automated software quality management and testing.[3] It provides a consistent repeatable process for gathering requirements; planning and scheduling tests; analysing results; and managing defects. It employs dashboard technology to give visibility into the process, and the tool consists of four modules, namely

- Requirements
- Test plan
- Test lab
- Defect management.

The Requirements module supports requirements management and traceability of the test cases to the requirements. The test plan module supports the creation and update of test cases. The test lab module supports execution of the test cases defined in the test plan module. The defect management module supports the logging of defects and these defects can be linked back to the test cases that failed.

HP Quality Center supports a high-level of collaboration and communication between the stakeholders. It allows the business analysts to define the application requirements and testing objectives. The test managers and testers may then design

[3]The tool was previously developed by Mercury and was called Test Director. Mercury is now part of HP.

test plans, test cases, and automated scripts. The testers then run the manual and automated tests, log the defects and report the results. The developers review and correct the logged defects. Project and test managers can create status reports and manage test resources. Test and product managers decide objectively whether the application is ready to be released.

TestRail is a Web-based test management tool that tracks, manages, and organizes the testing. It was created by Gurock (a subsidiary of US-based IDERA, Inc.), and the company produces various test tools for QA and development teams. The TestRail tool makes it easy to create test cases, manage test runs, and coordinate the testing process. It uses dashboards to track the status of individual tests, milestones, and the project.

QTest is a comprehensive test management tool that is often used in the Agile world, and it allows the testing to be organized so that testers can create and execute tests in different locations. It is a SaaS cloud-based tool that was created by QASymphony. There is no software to install as it is cloud-based and the project team may be in diverse geographical locations.

Once the project is created in qTest, and the start/end dates are added the navigation then displays test plan, requirements, test design, test execution, defects, and reports. QTest provides traceability between the requirements and defects, and it includes a test case repository to create, manage, and organize test cases. The test execution involves managing the test cycles and executing the tests and recording results. It provides defect-tracking and customisable reports, and it may be integrated (if desired) with Jira (qTest also has its own defect module for defect tracking).

QA Complete is a test management tool that provides traceability between the requirements and the test cases and defects. It may be integrated with other test tools such as Jira and Bugzilla, and it provides test coverage to ensure that test cases exist for all of the requirements.

### 10.2.1 Estimation and Scheduling Tools

There are several tools to support the various project management activities such as estimation and cost prediction, planning and scheduling, monitoring risks and issues, and managing a portfolio of projects. These include tools such as Microsoft Project (Fig. 5.2), which is a powerful project planning and scheduling tool. Small projects may employ a simpler tool such as Microsoft Excel for their project scheduling activities.

The Constructive Cost Model (Cocomo) is a cost prediction model developed by Boehm (1981), which is used to estimate effort, schedule, and cost for small and medium projects. It is based on an effort estimation equation that calculates the software development effort in person months from the estimated project size. The effort estimation calculation is based on the estimate of a project's size in thousands of *source lines of code* (SLOC[4]). The accuracy of the tool is limited.

Microsoft Project enables a realistic project schedule to be created, and the schedule is updated regularly during the project to reflect the actual progress made, and the project is re-planned as appropriate. The project schedule shows the tasks and activities to be carried out during testing; the effort and duration of each task and activity; the percentage complete of each task, and the resources needed to carry out the various tasks. The schedule shows how the testing will be completed within the key project parameters such as time, cost, and functionality without compromising quality in any way.

We discussed project management for testing in Chap. 5, and the test manager is responsible for preparing and managing the test schedule, and for taking corrective action when progress deviates from expectations. The project test schedule will be updated regularly to reflect the actual progress made, and the project is re-planned appropriately.

The test manager may employ tools for recording and managing risks and issues, and this may be as simple as using an excel spreadsheet. The test reporting may be done with a tool or with a standard Microsoft word report. Next, we consider a selection of tools for static testing.

## 10.3 Static Code Analysis Tools

Static code analysis is the analysis of software code without the actual execution of the code. It is usually performed with automated tools, and the actual analysis carried out depends on the sophistication of the tool. Some tools may analyse individual statements or declarations, whereas others may analyse the whole source code. The objective is to identify potential coding errors early in the software development lifecycle.

Some static code analysis tools (e.g., tools for formal methods) aim to prove properties about a particular program. This may include reasoning about program correctness or that of a program meeting its specification. These tools often provide support for assertions, where a precondition is the assertion placed before the code fragment, and this predicate must be true before execution of the code. The post-condition is the assertion placed after the code fragment, and this predicate must be true after the execution of the code.

There are several open-source tools available for static code analysis, and these include the RATS tools which provide multi-language support for C, C ++, Perl, and PHP, and the PMD tool for Java. There are several commercial tools available, and these include the LDRA Testbed tool which provides support for C, C ++, and Java; The Fortify tool helps developers to identify security vulnerabilities in C, C +

[4]SLOC includes delivered source lines of code created by project staff (excluding automated code generated and also code comments).

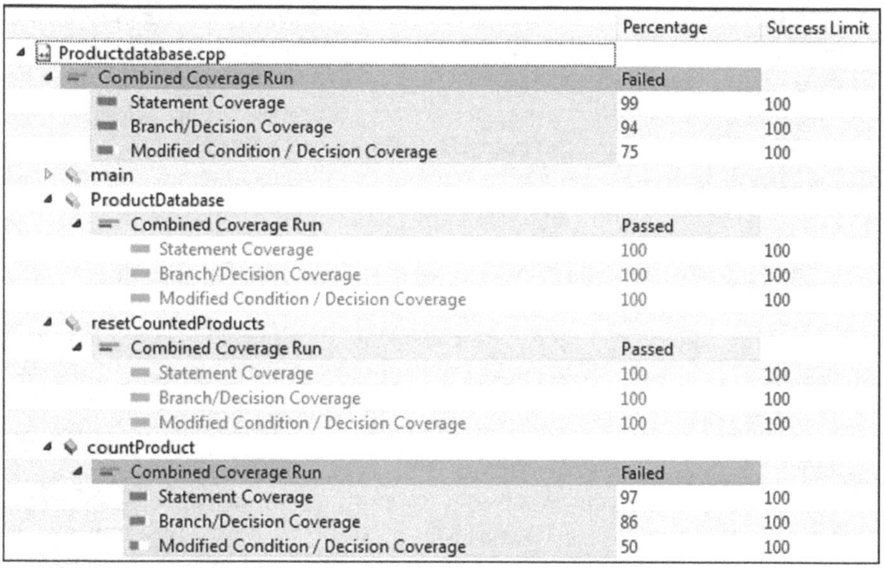

| | Percentage | Success Limit |
|---|---|---|
| Productdatabase.cpp | | |
| Combined Coverage Run | Failed | |
| Statement Coverage | 99 | 100 |
| Branch/Decision Coverage | 94 | 100 |
| Modified Condition / Decision Coverage | 75 | 100 |
| main | | |
| ProductDatabase | | |
| Combined Coverage Run | Passed | |
| Statement Coverage | 100 | 100 |
| Branch/Decision Coverage | 100 | 100 |
| Modified Condition / Decision Coverage | 100 | 100 |
| resetCountedProducts | | |
| Combined Coverage Run | Passed | |
| Statement Coverage | 100 | 100 |
| Branch/Decision Coverage | 100 | 100 |
| Modified Condition / Decision Coverage | 100 | 100 |
| countProduct | | |
| Combined Coverage Run | Failed | |
| Statement Coverage | 97 | 100 |
| Branch/Decision Coverage | 86 | 100 |
| Modified Condition / Decision Coverage | 50 | 100 |

**Fig. 10.2** LDRA Code Coverage Analysis Report

+, and Java; and the Parasoft tools helps developers to identify coding issues that lead to security, reliability, performance, and maintainability issues later. Among the tools discussed in this section include

- LDRA tools
- Jtest
- IBM rational software analyser
- CodeSonar.

The LDRA Tools automatically determine the complexity of the source code and provide metrics that indicate its maintainability. It gives a visual picture of system complexity, and it has a re-factoring tool to assist with its reduction. It generates code assessment reports listing all of the files examined and provides metrics of the clarity, maintainability, and testability of the code. Other LDRA tools may be used for code coverage analysis (Fig. 10.2).

Compliance to coding standards is important in producing readable code and in preventing error-prone coding styles. There are several tools available to check conformance to coding standards including the LDRA TBvision tool, which has reporting capabilities to show code quality as well as fault detection and avoidance measures. It provides functionality to view the results in various graphs and reports.

Jtest is an automated Java software testing and static analysis tool that was developed by Parasoft. It is used for static analysis, unit test case generation and execution, code coverage, run-time error detection, and regression testing. It audits the Java software code and produces a set of metrics of the code structure. It provides visibility into the compliance of the built-in static analysis rules for Java.

IBM Rational Software Analyser is a software analysis tool that is used to review code, identify defects, and to enforce coding standards early in the software development lifecycle.

CodeSonar is a static analysis tool for source code and binary executables, which was developed by GrammaTech in New York. It allows C, C ++, and Java source code and binary executables to be analysed, and it identifies software defects and security vulnerabilities in the software.

Next, we will consider a selection of tools to support test design including tools that derive test cases from the requirements.

## 10.4 Requirements and Test Design Tools

Test design tools are concerned with deriving the test cases from the specification of the requirements and design so that testing activities may be carried out. These tools analyse the requirements and design and assist in creating high-level test cases, generating test input and the desired output, ensuring that the test cases cover all of the requirements, and ensuring that the test cases cover all branches and paths of the software code.

These tools generally include functionality to manage changes to the requirements and allow the test cases to be maintained throughout the development lifecycle.

A requirements management tool can store the requirements and identify undefined or missing requirements. They provide traceability to the design and test cases and usually interface with various test management tools to ensure requirements coverage during testing. There are several tools available to support requirements management (Table 10.4). These assist in defining the requirements and managing them throughout the project lifecycle and provide traceability to the design and test cases.

DOORS® (Dynamic Object-Oriented Requirements System) is a requirements management tool developed by IBM Rational (Fig. 10.3). It aims to optimize requirements communication, collaboration and verification in order to achieve business objectives.[5]

The tool can capture, link, trace, analyse, and manage changes to the requirements. Requirements are documented in a way that is easy to interpret and navigate. The user requirements are recorded in a document style showing each individual requirement in an Explorer-like navigation tree.

[5]A good requirements process will enable high-quality requirements to be consistently produced, and minimize wastage and rework. The requirements are the foundation of the system, and if they are incorrect, then the delivered system will be incorrect.

**Table 10.4** Tools for requirements development and management

| Tool | Description |
|---|---|
| Doors (IBM/Rational) | This is a requirements management tool developed by Telelogic (which is now part of IBM/Rational) |
| Enterprise Architect (Sparx Systems) | This is a UML analysis and design tool that covers requirements gathering, analysis and design, and testing and maintenance. It was developed by Sparx Systems and integrates requirements management with the other software development activities |
| Core (Vitech) | This is a requirements tool developed by Vitech, which may be used for modelling and simulation |
| Integrity (MKS) | This tool was developed by MKS and enables organizations to capture and validate software requirements and to link them to downstream development and testing activities |

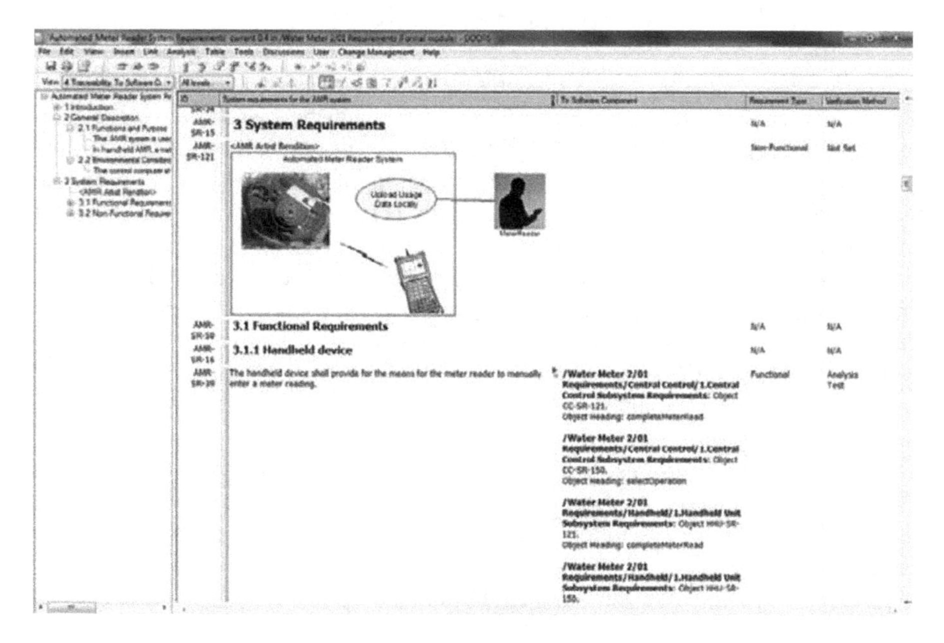


**Fig. 10.3** IBM Rational Doors Tool

The tool employs links to support traceability of the requirements, and these are traversed with a simple click of the mouse to the corresponding object. The links are easy to create by dragging and dropping: e.g., a new link from the user requirements to the system requirements is created in this way.

Doors supports changes to the requirements with an impact analysis of the proposed changes performed. It allows changes that could impact other requirements or design items and test cases to be tagged. The IBM Rational System Architect, Requirements Composer, Rhapsody, and Quality Manager may be integrated with Doors.

The Core product was developed by Vitech and is used for requirements management, modelling, and simulation, and for verification and validation. It supports UML activity and sequence diagrams, which describe the desired behaviour and flow of control. The tool provides comprehensive end-to-end system traceability and impact analysis of changes, as well as the generation of documentation from the database.

The Integrity tool was developed by MKS, and it enables organizations to capture and validate software requirements. It enables them to link the requirements to downstream development and testing activities and to manage changes to the requirements.

Test design tools analyse the specifications and requirements and may create high-level test cases or generate test input. The tools help to reduce the time and effort involved in specifying the test cases. The various test management tools discussed earlier in Sect. 10.2 may also be used in test design.

QASymphony's cloud-based qTest tool is used in the Agile world, and it allows the testing to be organized so that testers can create and execute tests in different physical locations. It includes a test case repository to create and manage test cases and provides traceability between requirements and defects.

QAComplete is a comprehensive test management tool that allows manual and automated tests to be linked to requirements and defects. It provides traceability between the requirements, test cases and defects, and determines test coverage and that tests exist for all requirements.

TestLink is an open-source test management tool, and this Web-based tool includes planning, reporting, test specification, and requirements tracking. It provides charts and reports to give visibility into the progress with the testing, and it allows defects to be logged and metrics to be generated. Next, we consider test execution tools.

## 10.5 Test Execution Tools

A test execution tool is basically a tool that can run tests during the testing phase, and they include tools to capture/playback tests, tools to debug the code, tools to simulate parts of the operational environment, security tools, and tools to provide visibility on statement and branch coverage:

- capture/playback
- debugging tools
- test harness tools
- security tools
- test coverage tools.

Capture/playback tools operate by capturing or recording manual tests which they can then playback, and they employ a scripting language to run the tool. The scripting language is similar to a programming language, and the software tester creates and modifies the test scripts.

These tools record the test inputs while the tests are executed manually and store an expected result to compare to the next time the test is run. The tool may execute the test scripts and do a comparison and log the results of testing from a comparison of expected/actual results.

The advantage of the scripting language is that tests can repeat actions for different test inputs, and they can take a different route depending on the result of a test. That is, if the test fails, it may go to a different set of tests from the set of tests that it would go to if it passed. The advantage of capture/playback tools is that they improve test productivity, but their disadvantage is that a small change in the software can invalidate several (many) of the scripts. This is especially the case when the GUI structure is changed, as the test cases written for the old GUI often become obsolete. Capture/playback tools are employed in regression testing to improve test productivity.

*Debugging tools* are used by the software developers to step through the software code to localize and fix a software defect (e.g., a memory leak problem or unassigned pointers). One of the earliest debuggers was CodeView which was developed for the MS/DOS operating system platform in the mid-1980 s. It was a full-screen debugger that presented the user with several windows (such as a code window, a data window, and a watch window). Today, there are a plethora of debugging tools available for different platforms such as Windows 10, Linux, and Java.

*Test harness tools* are used mainly by the software developers to simulate the operational environment, and they provide the stubs and driver programs that interact with the software. The stubs and drivers are used to replace missing or unavailable parts of the system, and they provide the required information to the software, and they receive any information sent by the software. Test execution with the test harness involves executing a suite of tests, providing the test input to the software, receiving the output from the software and recording whether the test passes or fails.

JUnit is a unit testing framework for the Java programming language, and it plays an important role in test driven development frameworks. It is a member of the xUnit family of unit testing frameworks. Jtest includes functionality for unit test case generation and execution, as well as static testing.

*Test coverage tools* are used to calculate the percentage of coverage items (e.g., statements, decisions, function calls, etc.) that are tested by a suite of tests (e.g., the LDRA test coverage tools discussed in Sect. 10.3). They include features to identify the coverage items, the calculation of the percentage coverage, and the reporting of the results.

*Security tools* are used to test the security of a system by identifying viruses or attempting to hack into it. The objective is to reveal flaws in the security mechanisms of a computer system that protect data and functionality. There are several security requirements such as confidentiality, availability, and integrity of information.

These include tools such as Metasploit which allows security assessments to be performed to identify system vulnerabilities and to improve awareness of the importance of system security. Wireshark is a popular open-source packet analyser capable of providing users with detailed information on network protocols and packet information.

### 10.5.1 Tools for Regression Testing

Capture/playback tools replay tests that have previously been manually run, and so they are very suitable for regression testing which aims to verify that the functionality of the existing system has not been compromised following changes to the software. Changes may be due to the correction of a defect or an enhancement to the software, and it is essential to verify that no new defects have been introduced. There are many regression tools available including

- WinRunner (HP Unified Testing Software)
- HP Quick Test Professional (QTP)
- TestingWhiz
- Selenium
- Borland Silk Test
- IBM Rational Functional Tester.

Mercury developed the WinRunner tool that automatically captures, verifies, and replays user interactions. It is used mainly used to automate regression testing, which improves test productivity and provides confidence that changes to the software have preserved the integrity of the system. The tool has been replaced by HP Unified Functional Testing Software, which includes HP Quick Test Professional and HP Service Test.

The HP Unified Functional Testing (UFT) tool provides functional and regression test automation for software applications, and it uses VB script as the scripting language to specify the test procedure to manipulate the objects. It works by identifying the objects in the user interface or Web page and performs the desired operations (such as mouse clicks or keyboard entry).

TestingWhiz is a test automation tool for Web, mobile, and cloud applications. Testers can schedule the tests to run at a specified time and can also execute them when feasible. It has in-built capture/playback functionality that may be used to regularly test the system following continuous integrations to verify that the core system is functioning correctly and that no new defects have been introduced.

Selenium is an open-source software testing framework for Web applications, and it provides a playback feature for authoring tests without the need to learn a scripting language. It also provides a specific scripting language (Selenese) to write tests in several popular programming languages such as Java and Perl.

Borland Silk Test is a tool to automate functional and regression testing for software applications. It was originally developed by Segue Systems (which later became part of Borland), and Borland is now part of Micro Focus. It uses VB.net as its scripting language.

IBM Rational Functional Tester is a tool for the automated testing of software applications. The tool automates functional and regression testing, and it allows users to create tests that mimic the action and assessments of a human tester. It enables testers to automate tests by recording user actions as well as providing customization options.

## 10.6 Tools for Defect Tracking

Jira is a popular tool that was developed Australian company, Atlassian in 2002, and it was initially used just for defect tracking and issue tracking. The current version of Jira includes packages that enable it to be used as an IT service desk or as a generic project management tool.

The PV Tracker tool automates the capture and communication of issues and change requests. This is done throughout the software development lifecycle for project teams, and the tool allows the developers to link the affected source code files to issues and changes. It allows managers to determine and report on team progress and to prioritise tasks. PV Builder maintains an audit trail of the files included in the build as well as their versions.

Bugzilla is a Web-based general-purpose defect tracking tool that was originally developed in the late 1990s for the Mozilla project (Fig. 10.4). It is licensed under the Mozilla public license. Netscape released it as open-source software in the late

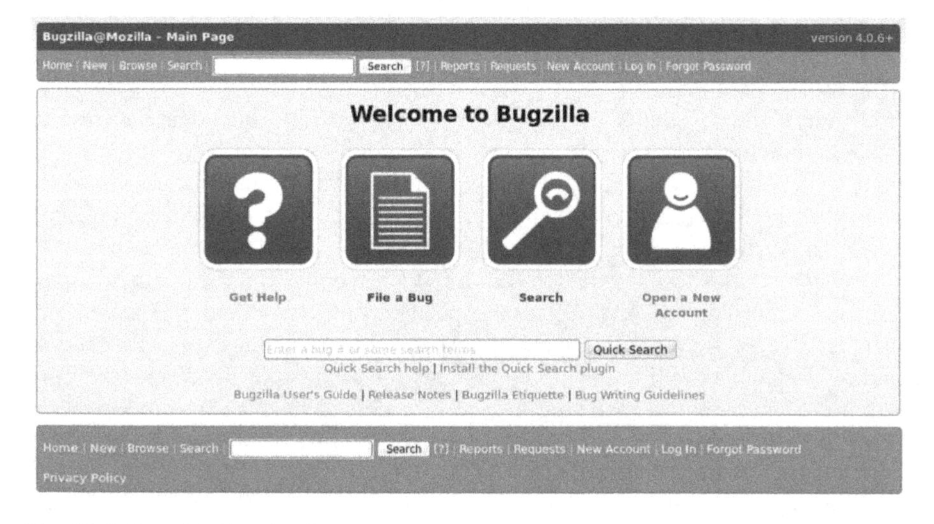


**Fig. 10.4** Bugzilla: Creative Commons

1990 s, and it has been adapted for use as a bug tracking system by many organizations.

IBM Rational ClearQuest allows the defects in a project to be tracked, and it enables the versions of source code modules to be changed to be linked to a defect number in ClearQuest.

## 10.7 Test Performance and Monitoring Tools

The purpose of performance testing is to determine how the system will respond to a certain level of utilization, where the users could potentially be in different geographical locations around the world using different networks, browsers, and devices. The objective is to determine how the system responds to the expected load and to stressful loads, and to determine how many concurrent users the Web site can handle, and the response time for a given number of concurrent users.

The performance test tools generally record the test scripts, and these may then be replayed. A test control language is employed, and the test data needs to be stored and controlled. There are several tools available for performance testing including:

- HP LoadRunner
- Borland Silk Performer
- IBM Rational Performance Tester
- Apache JMeter.

Mercury originally developed the LoadRunner performance-testing tool, which allows the performance of a software application to be tested by simulating thousands of concurrent users. It allows the scalability of the software system to be determined, and the extent to which the application can support the future predicted growth is investigated by measuring its performance under heavy system loads. The LoadRunner tool is now part of Micro Focus, which acquired the Mercury test tools.

Borland Silk Performer (part of Micro Focus) is a performance-testing tool that is designed to deliver a consistent user experience. It may be used to predict and prevent outages, and it may simulate users in a cloud environment. Segue Systems originally developed it.

IBM Rational Performance Tester validates the scalability of Web and server applications. It identifies the presence and causes of system performance bottlenecks and allows testers to execute performance tests to analyse the impact of loads on applications.

Apache JMeter is an Apache project tool that can be used as a load testing tool for analysing and measuring system performance (Fig. 10.5).

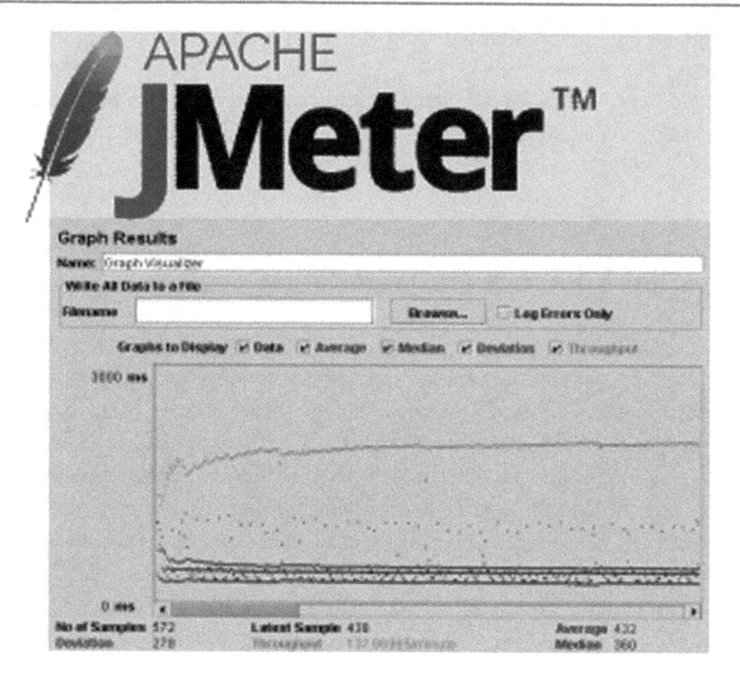


**Fig. 10.5** Apache JMeter: Creative Commons

## 10.8 Tools for Testing in Agile World

Testing is an essential part of the Agile world, and testers and developers on Agile projects need effective tools to manage the testing and to report the defects identified. These include test automation and test management tools including:

- BugDigger
- QTrace
- Usersnap
- Snagit.

BugDigger can be used during testing to create and submit defect reports to popular defect-tracking and project management tools such as Jira. It allows screenshots of the system to be taken as well as including additional context information relevant to the defect report.

QTrace is a screen capturing tool that allows screenshots to be annotated, as well as allowing the steps to be entered to reproduce the issue, which may then be submitted as a defect report to popular defect tracking tools.

Usersnap is a visual bug tracking and feedback tool that allows testers and end users to generate defect reports for the Web site or Web application. The tool can be integrated with defect tracking tools such as Jira, and it was developed by an Austrian company. Snagit is a popular screen capturing tool. The next section is concerned with tools to support configuration management.

## 10.9 Tools for Configuration Management

Configuration management is concerned with identifying the work products that are subject to change control, and controlling changes to them. It involves creating and releasing baselines, maintaining their integrity, recording and reporting the status of the configuration items and change requests, and verifying the correctness and completeness of the configuration items with configuration audits.

A version control management system for source code and binary files is used mainly by development organizations to place their source code and work products under version control management.

Polytron Version Control System (PVCS) is a version control system for software code and binary files. It was developed by Serena Software Inc. and is suitable for large or small teams. It allows multiple users to place their source code and project deliverables under version control management, and it allows files to be checked in and checked out; baselines to be controlled; rollback of code; and tracking of check-ins and checkouts. It includes functionality for branching, merging, and labelling. It includes the PV Tracker tool for tracking defects, and the PV Builder tool for performing builds and releases.

IBM Rational ClearCase is a popular configuration management tool with a rich feature set. It allows software code and other software deliverables to be placed under version control management, and it may be employed in large or medium projects. It can handle a large number of files and supports standard configuration management tasks such as checking in and checking out of the software assets as well as labelling and branching.

## 10.10 Review Questions

1. Why types of tools are used in software testing?
2. How should a tool be selected?
3. What is the relationship between the process and the tool?
4. What tools would you recommend for test management? Why?
5. Describe how you would go about selecting a tool for test execution.

6. Describe various tools that are available for defect tracking.
7. What tools would you recommend for performance testing? Why?
8. What tools would you recommend for configuration management?

## 10.11 Summary

The objective of this chapter was to give a flavour of some of the tools available to support testing in the organization. These included tools for the management of testing, tools for the design and execution of the testing, tools for static testing, tools for regression testing, defect tracking tools, performance-testing tools, and configuration management tools. The tools are generally chosen to support the process rather than adjusting the process to support the tool.

Tool selection is best done in a controlled manner. First, the organization needs to determine its requirements for the tool. Various candidate tools are evaluated, and a decision is made on the proposed tool. Next, the tool is piloted to ensure that it meets the needs of the organization, and feedback from the pilot may lead to changes or customizations of the tool. Finally, the end users are trained on the use of the tool and it is rolled out throughout the organization.

We discussed several test management tools including HP Quality Center™, which standardizes and manages the entire test process. It has modules for requirements management, test planning, test lab, and defect management. We briefly discussed cost estimating and scheduling, and mentioned tools such as Microsoft Project tool for test scheduling, and the Cocomo cost estimator model.

We discussed tools to support static testing including the LDRA tools, which provide reports on code complexity and compliance to coding standards. We discussed capture/playback tools for regression testing including WinRunner and Selenium. We discussed tools for performance testing including LoadRunner and Silk Performer tools.

We discussed tools for defect tracking including Jira and Bugzilla. Finally, we discussed tools to support configuration management, including PVCS and ClearCase.

## Reference

Boehm B (1981) Software engineering economics. Prentice Hall, New Jersey

# 11 Test Process Improvement

**Key Topics**

Software process
Software process improvement
Process mapping
Benefits of software process improvement
CMMI
ISO/IEC 15504 (SPICE)
ISO 9000
PSP and TSP
TPI model
TMM model, CTP, STEP and TPI
PDCA, IDEAL
Verification and validation
Root cause analysis

## 11.1 Introduction

The success of business today is highly influenced by the functionality and quality of the software that it uses. It is essential that the software is safe, reliable, of high quality, and fit for purpose. Companies may develop their own software internally, or they may acquire software solutions off the shelf or from bespoke software development. Software development companies need to deliver high-quality and reliable software consistently on time to their customers.



G. O'Regan, *Concise Guide to Software Testing*,
Undergraduate Topics in Computer Science,
https://doi.org/10.1007/978-3-030-28494-7_11

Cost is a key driver in most organizations and it is essential that software is produced as cheaply and efficiently as possible, and that waste is reduced or eliminated in the software development process. In a nutshell, companies need to produce software that is *better, faster, and cheaper* than their competitors in order to survive in the market place. They need to continuously work smarter to improve their businesses and to deliver superior solutions to their customers.

Software process improvement initiatives play a key role in helping companies achieve their key goals. They assist in the implementation of best practice in organizations and allow companies to focus on fire prevention rather than fire-fighting. They allow companies to solve key issues to eliminate quality problems and to critically examine their current processes to determine the extent to which they are meeting their needs, as well as identifying how they may be improved to eliminate inefficiencies.

Software process improvement allows companies to identify the root causes of problems and to determine appropriate solutions. Its benefits include the consistent delivery of high-quality software, improved financial results, and increased customer satisfaction.

The focus in software process improvement is on the process and on ways to improve it, as problems are often caused by a defective process rather than people. Further, a focus on the process helps to avoid the blame culture that arises when blame is apportioned to individuals rather than the process. The focus on the process leads to a culture of openness in discussing problems and their solutions.

This chapter is concerned with test process improvements, and we shall discuss several test maturity models (e.g., TMM*i*, TMap, TPI, STEP and CTP) that are useful in software testing. First, we introduce the wider software process improvement field and discuss the nature of a software process.

## 11.2 Software Process Improvement

The origins of the software process improvement field go back to Walter Shewhart's work on statistical process control in the 1930s (see Chap. 1). Shewhart's work was later refined by Deming and Juran, who argued that high-quality processes are essential to the delivery of a high-quality product. They argued that the quality of the end product is largely determined by the processes used to produce and support it and that there needs to be an emphasis on the process as well as on the product.

Deming argued that product quality will improve as variability in process performance is reduced, and his approach was effective in transforming manufacturing companies with quality problems to companies that would consistently deliver high-quality products. Further, the improvements to quality led to cost reductions and higher productivity, as less time was spent in reworking defective products.

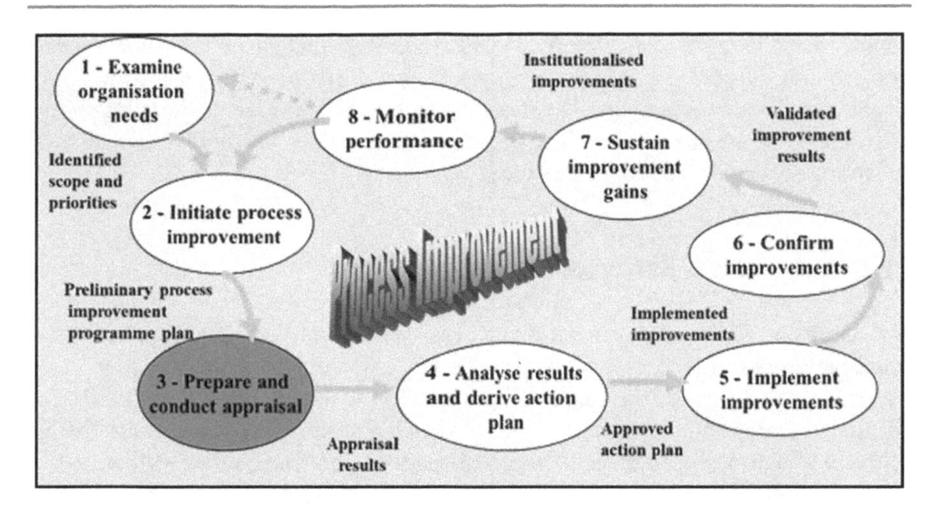


**Fig. 11.1** Steps in process improvement

Watt Humphrey and others at the SEI applied Deming and Juran's approach to the software quality field leading to the birth of the software process improvement field (SPI). Software process improvement is concerned with practical action to improve the software processes in the organization to ensure that business goals are achieved more effectively (Fig. 11.1).

**Definition 11.1** (*Software Process Improvement*) A programme of activities designed to improve the performance and maturity of the organization's software processes and the results of such a program.

Software process improvement initiatives support the organization in achieving its key business goals more effectively, where the business goals could be delivering software faster to the market, improving quality, and reducing or eliminating waste. The objective is to work smarter and to build software better, faster, and cheaper than competitors. It makes business sense and provides a return on investment.

There are international standards and models available to support software process improvement such as the CMMI Model, the ISO 90001 standard, and ISO 15504 (popularly known as SPICE). The CMMI model includes best practice for processes in software and systems engineering. The ISO 9001 standard is a quality management system that may be employed in hardware, software development, or service companies. The ISO 15504 standard is an international standard for software process improvement and process assessment, which is popular in the automotive and medical device sectors.

Software process improvement is concerned with defining the right processes and following them consistently. It involves training all staff on the new processes, refining the processes, and continuously improving them. The need for a process improvement initiative often arises due to the realization that the organization is weak in some areas in software engineering, and that it needs to improve to achieve its business goals more effectively. The starting point of any improvement initiative is an examination of the business needs of the organization, such as improving quality or delivering products faster to the market.

### 11.2.1 What Is a Software Process?

A software development process is the process used by software engineers to design and develop computer software. It may be an undocumented ad hoc process as devised by the team for a particular project, or it may be a standardized and documented process used by various teams on similar projects. The process is seen as the glue that ties people, technology and procedures coherently together.

The processes employed in software development include processes to determine the requirements; processes for the design and development of the software; processes to verify that the software is fit for purpose; and processes to maintain the software.

A *software process* is a set of activities, methods, practices and transformations that people use to develop and maintain software and the associated work products.

**Definition 11.2** (*Software Process*) A *process* is a set of practices or tasks performed to achieve a given purpose. It may include tools, methods, material and people.

The process is an abstraction of the way in which work is done in the organization, and it is seen as the glue (Fig. 11.2) that ties people, procedures and tools together. An organization will typically have many processes in place for doing its work, and the object of process improvement is to improve these to meet business goals more effectively.

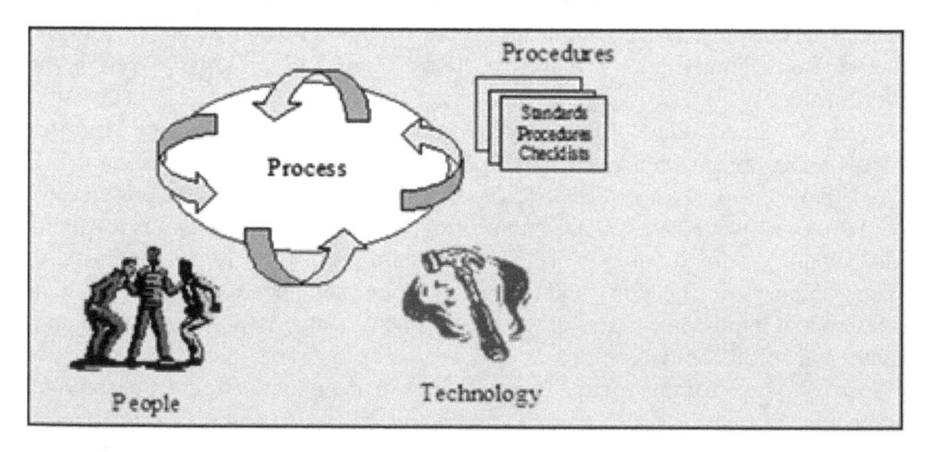


**Fig. 11.2** Process as glue for people, procedures and tools

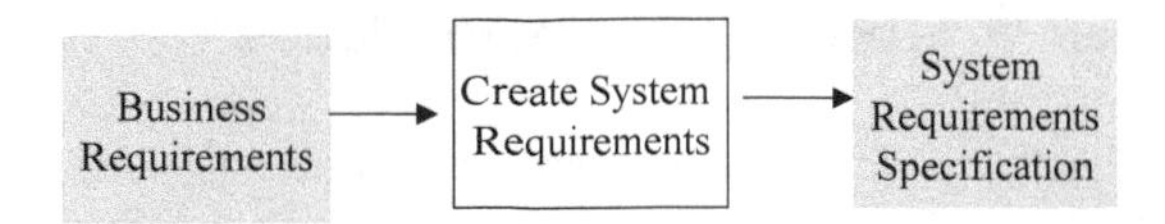


**Fig. 11.3** Sample process map

The Software Engineering Institute (SEI) argues that there is a close relationship between the quality of the delivered software and the quality and maturity of the underlying processes employed to create the software. It developed process maturity models such as the CMM and its successor the CMMI. These maturity models are invaluable in maturing the software processes in software-intensive organizations.

A process is often represented by a process map which details the flow of activities and tasks. The process map will typically include the inputs to each activity as well as the output from an activity. Often, the output from one activity will become an input to the next activity. A simple example of a process map for creating the system requirements specification is described in Fig. 11.3.

As a process matures, it is defined in more detail and documented. It will have clearly defined entry and exit criteria, inputs and outputs, an explicit description of the tasks, verification of the process and consistent implementation throughout the organization.

### 11.2.2 Benefits of Software Process Improvement

Often projects encounter problems such as budget and schedule overruns, late delivery of the software, spiralling costs, quality problems, and customer complaints. Software process improvement helps to deal with these challenges and its benefits include:

- Improvements to customer satisfaction
- Improvements to on-time delivery
- Improved consistency in budget and schedule delivery
- Improvements to quality
- Reductions in the cost of poor quality
- Improvements in productivity
- Reductions to the cost of software development
- Improvements to employee morale.

### 11.2.3 Software Process Improvement Models

A process model[1] such as the CMMI defines best practice for software processes in the organization. It describes what the processes should do rather than how they should be done, which allows professional judgment to be used in the implementation of processes. The process model will need to be interpreted and tailored to meet the needs of the organization.

A process model provides a place to start an improvement initiative, and it provides a common language and shared vision for improvement. It provides a framework to prioritize actions, and it allows the benefits of the experience of other organizations to be shared. The popular process models used in software process improvement include:

- Capability Maturity Model Integration (CMMI)
- ISO 9001 Standard
- ISO 15504
- PSP and TSP

The CMMI provides a structured approach to improvement, which allows the organization to set its improvement goals and priorities. It provides a clearly defined roadmap for improvement, and it allows the organization to improve at its own pace. Its approach is evolutionary rather than revolutionary, and it recognizes that a balance is required between project needs and process improvement needs. It allows the processes to evolve from ad hoc immature activities to disciplined mature processes.

A SCAMPI appraisal determines the actual process maturity of an organization, and a SCAMPI class A appraisal allows the organization to benchmark itself against other organizations.

ISO 9001 is an internationally recognized quality management standard (Fig. 11.4), and it is customer and process focused. It applies to the processes that an organization uses to create and control products and services, and it emphasizes continuous improvement[2]. The standard is designed to apply to any product or service that an organization supplies.

The ISO/IEC 15504 standard (popularly known as ISO SPICE) is an international standard for process assessment. It includes guidance for process improvement and for process capability determination, as well as guidance for performing an assessment. It uses the international standard for software and systems lifecycle processes (ISO/IEC 12207) as its process model.

The Personal Software Process (PSP) is a disciplined data-driven process that is designed to help software engineers understand and to improve their personal software process performance. It helps engineers to improve their estimation and planning skills and to reduce the number of defects in their work. This enables them

[1]There is the well-known adage "All models are wrong, some are useful".

[2]The ISO 9004 standard provides guidance on continuous improvement.

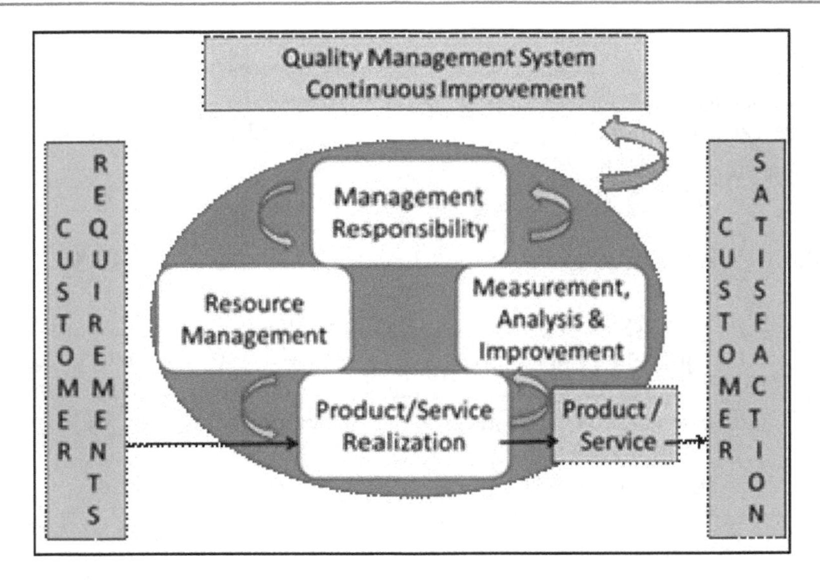


**Fig. 11.4** ISO 9001 quality management system

to make commitments that they can keep and to manage the quality of their projects.

The Team Software Process (TSP) is a structured approach to help software teams understand and improve their quality and productivity. Its focus is on building an effective software development team, and it involves establishing team goals, assigning team roles as well as other teamwork activities. The team members must already be familiar with the PSP.

### 11.2.4 Process Mapping

The starting point for improving a process is to first understand the process as it is currently performed and to then determine its effectiveness. The stakeholders reach a common understanding of how the process is currently performed, and the process is then sketched pictorially, with the activities and their inputs and outputs recorded graphically. This graphical representation is termed a "*process map*" and is an abstract description of the process "*as is*".

The process map is an abstraction of the way that work is done, and it is critically examined to determine its effectiveness and to identify weaknesses and potential improvements. This leads to modifications to the process, and the proposed new process is sketched in a new process map to yield the process "*to be*".

Each activity has an input and an output, and these are recorded in the process map. Once the definition of the new process is agreed, the supporting templates are identified from an examination of the input and output of the various activities. There may be a need for standards to support the process (e.g., procedures and templates), and the procedures or guidelines provide the details on how the process is to be carried out, and they will detail the tasks and activities, and the roles required to perform them.

### 11.2.5 Process Improvement Initiatives

The need for a software process improvement initiative often arises from the realization that the organization is weak in some areas in software engineering, and that it needs to improve to achieve its business goals more effectively. The starting point is an examination of the business goals of the organization such as:

- Delivering high-quality products on time
- Delivering products faster to the market
- Reducing the cost of software development
- Improving software quality.

There is more than one approach to the implementation of an improvement programme. A small organization has fewer resources available and team members involved in the initiative will typically be working part time. Larger organizations may be able to assign some people full time to the improvement activities.

Once the business goals have been defined, the improvement initiative commences. This involves conducting an appraisal to determine the current strengths and weaknesses of the processes; analyzing the results to formulate a process improvement plan; implementing the plan; piloting the improved processes and verifying that they are effective; training staff and rolling out the new processes (Fig. 11.5). The improvements are monitored for effectiveness and the cycle repeats. The software process improvement philosophy is:

- The improvement initiative is based on business needs
- Improvements are based on the strengths and weaknesses of the processes
- The improvements are prioritized (it is not possible to do everything at once)
- The improvement initiative needs to be planned and managed as a project
- The results achieved are reviewed at the end of cycle, and a new improvement cycle started
- Organization culture (and training) needs to be considered
- There needs to be a process champion/project manager
- Senior management needs to be 100% committed to the success of the initiative
- Staff need to be involved in the improvement initiative, and there needs to be a balance between project needs and the improvement activities.

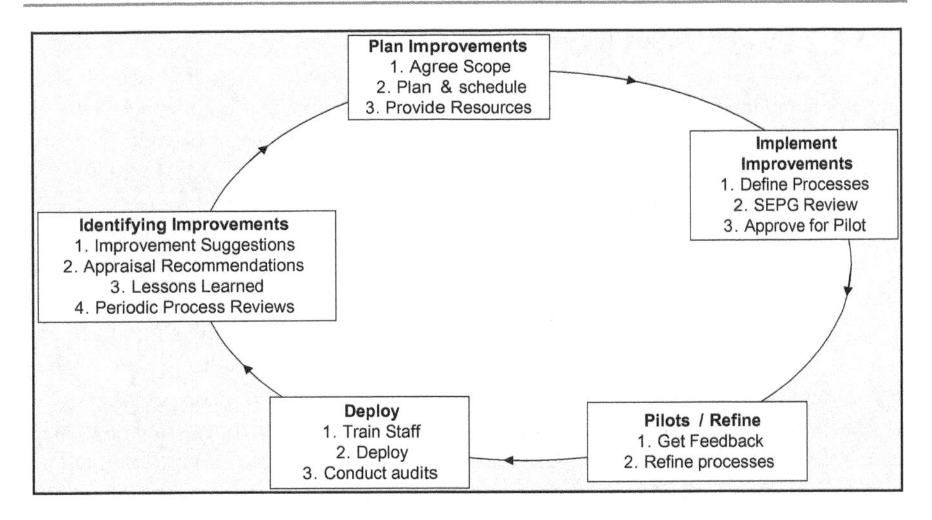


**Fig. 11.5** Continuous improvement cycle

### 11.2.6 Barriers to Success

Software process improvement initiatives are not always successful and occasionally are abandoned. Some of the reasons for failure are:

- Unrealistic expectations
- Trying to do too much at once
- Lack of senior management sponsorship
- Focusing on a maturity level
- Poor project management of the initiative
- Insufficient involvement of staff
- Insufficient time to work on improvements
- Inadequate training on software process improvement
- Lack of pilots to validate new processes
- Inadequate training/rollout of new processes.

It is essential that a software process improvement initiative be treated as a standard project with a project manager assigned to manage the initiative. Senior management needs to be 100% committed to the success of the initiative, and they need to make staff available to work on the improvement activities. It needs to be clear to all staff that the improvement initiative is a priority to the organization. All employees need to receive appropriate training on software process improvement and on the process maturity model.

### 11.2.7 Setting Up an Improvement Initiative

The implementation of an improvement initiative is a project, and it needs to be managed as such. The project manager will prepare plans to implement the initiative within the approved schedule and budget. The project may consist of several improvement cycles, with each improvement cycle implementing one or more process areas.

One of the earliest activities carried out on any improvement initiative is to carry out an appraisal to determine the current strengths and weaknesses of the processes, as well as gaps with respect to the practices in the model. This allows management in the organization to understand its current maturity with respect to the model and to communicate where it wants to be, as well as how it plans to get there.

The project manager then prepares a project plan and schedule. The plan will detail the scope of the initiative, the budget, the process areas to be implemented, the teams and resources required, the initial risks identified, the key milestones, and so on. The schedule will detail the deliverables to be produced, the resources required, and the associated timeline for delivery.

The steps in the initiative include examining organization needs; conducting an appraisal to determine the current strengths and weaknesses; and analysing the results to formulate an improvement plan. The improvement plan is then implemented; the improvements monitored and confirmed as being effective; and the improvement cycle repeats. The steps in the improvement cycle are:

- Identify improvements to be made
- Plan improvements
- Implement improvements
- Pilots/refine[3]
- Deploy
- Do it all again.

### 11.2.8 Appraisals

Appraisals (Fig. 11.6) allow an organization to understand its current software process maturity. An initial appraisal is conducted at the start of the initiative to allow the organization to plan and prioritize improvements for the first improvement cycle. The improvements are then implemented, and an appraisal is conducted at the end of the cycle to confirm the progress made.

[3]The result from the pilot may be that the new process is not suitable to be deployed in the organization or that it needs to be significantly revised prior to deployment.

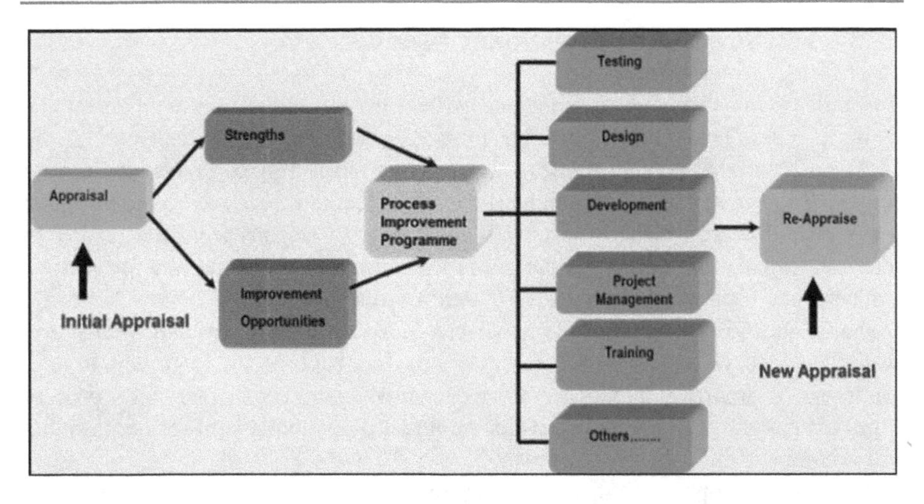


**Fig. 11.6** Appraisals

An appraisal is an independent examination of the software engineering and management practices in the organization, and it will identify strengths and weaknesses in the processes, and any gaps that exist with respect to the maturity model.

The appraisal activities include presentations, interviews, reviews of project documentation, and detailed analysis to determine the extent to which the practices in the model have been implemented. The appraisal leader will present the appraisal findings, and this may include a presentation and an appraisal report. The appraisal output summarizes the strengths and weaknesses, and ratings of the process areas will be provided (where this is part of the appraisal). The findings are used to plan the next improvement cycle, and it allows the organization. to:

- Understand its current process maturity (including strengths and weaknesses)
- Relate its strengths and weaknesses to the improvement model
- Prioritize its improvements for the next improvement cycle
- Benchmark itself against other organizations.

There are three phases in an appraisal

- Planning and preparation
- Conducting the appraisal
- Reporting the results.

## 11.3 Test Process Improvement Models

Test process improvement is concerned with the continuous improvement of the testing process. There are dedicated test process improvement models such as the TMM or TMap that contain best practice in software testing to mature the testing process. The test manager also conducts a lessons learned review with the test team at the end of a project, and the goal is to learn any lessons from the testing to determine what went well and what needs to be improved for the next project.

There has been a trend towards shortening time to market for new software products, and this has placed additional pressures on developers and testers in a deadline-driven culture. Process improvement has become an important tool to continuously improve processes to meet business needs, and test process improvement plays a key role in developing a test process that is fit for purpose that will continue to meet business needs.

Test process improvement offers a way to rigorously assess the maturity of the testing practices in an organization, and it provides an improvement roadmap as well as defining the steps that the organization needs to take to mature its test practices. There are various models to support test process improvement (Table 11.1).

Next, we describe these models in more detail.

**Table 11.1** Test process improvement models

| Model | Description |
|---|---|
| TMM Model | The Test Maturity Model is a framework to assess the maturity of test processes in the organization. It is complementary to the CMM, and it was originally developed by the Illinois Institute of Technology. It is now managed by the TMM*i* Foundation |
| TMap Model | The TMap model is a business-driven and risk-based approach to testing. It was developed by Sogeti (a subsidiary of Capgemini) |
| TPI Model | The TPI model is a framework to assess the maturity of the test practices in an organization. It was developed by Sogeti |
| STEP Model | Systematic Test Evaluation Process provides a process model for software testing. It was developed by Bill Hetzel and David Gelperin |
| CTP Model | Critical Test Process (CTP) is a lightweight framework for test process improvement. It was developed by Rex Black of RBCS |
| PDCA | The plan–do–check–act model (PDCA) is concerned with planning the improvements to be made; making the improvements; checking that the improvements are effective; and acting by analyzing the results and adjusting the process |
| CMMI | The Capability Maturity Model is used to implement best practice in software and system engineering, and it includes the verification and validation process areas for software testing |

### 11.3.1 TMM*i* Model

The *Test Maturity Model* (TMM) provides a framework to assess the maturity of the test practices in an organization, and it provides a roadmap for improvement. It is complementary to the CMMI, and relevant practices from the CMMI are referenced in the model. The TMM was originally developed by the Illinois Institute of Technology in the late 1990s, but it is now maintained by the TMM*i* foundation. The Test Maturity Model Integration (TMM*i*) is the successor to the historical TMM model.

The TMM*i* model provides a staged approach to improvement, and it consists of five maturity levels that an organization passes through as its testing process evolves from an ad hoc unmanaged process to one that is managed, defined, and optimized. Process improvement with the TMM*i* consists of moving through the different maturity levels, with each new level representing an increase in software testing capability, and acting as a foundation for the next level. There is a TMM*i* assessment framework (based on ISO 15504) that allows assessments (self-assessments or formal assessments) to be conducted. The five maturity levels in the TMM*i* model are summarized in Table 11.2.

Each maturity level consists of several process areas, and each process area includes goals and activities to implement the goals. The maturity level is achieved when all of the goals are satisfied, and the achievement of a higher maturity level represents an increase in test capability (Fig. 11.7).

The TMMi Assessment framework allows self-assessments or formal assessments of the testing process against the TMMi reference model. The assessment method includes activities such as preparation, conducting the assessment, reporting the results, analysing the results, and planning and implementing the improvements.

**Table 11.2** TMM*i* model

| Level | Description |
|---|---|
| Initial | Ad hoc/chaotic undefined testing practices employed and the results are not repeatable. Not a stable environment to support the testing process |
| Managed | Test strategy is in place and test plans are defined for each project (specifying the what/when/who of testing). Risk management and test monitoring and control of testing are in place |
| Defined | Testing is fully integrated into the development lifecycle and test planning done early in the project. There is an organization set of standard test processes that are tailored to individual projects, and a formal review program implemented. There is a dedicated trained test organization |
| Measured | An organization wide test measurement program is in place that is used to measure the quality of the testing process and to measure productivity and improvements |
| Optimization | The test process is continuously improved through incremental and innovative process and technology improvements (e.g., tools/defect prevention) |

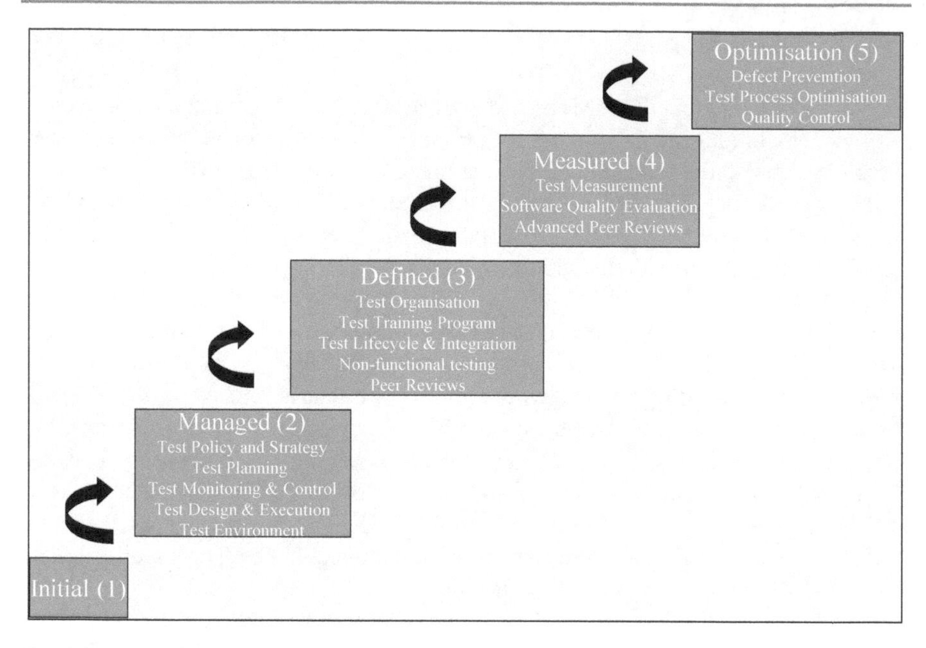


**Fig. 11.7** TMMi maturity levels

### 11.3.2 TMap Next Model

Martin Pol and others developed an early version of the *Test Management approach* (TMap) in the mid-1990s, and it was improved and developed further by Sogeti (a subsidiary of Capgemini). Tmap Next is the successor to the original TMap model, and it was introduced in 2006. It is a structured approach to testing that enables defects to be identified early as well as allowing test cases to be reused.

TMap is a business-driven and risk-based approach to testing, and its structured test process is described using the TMap lifecycle model. It contains lots of practical information to support the test process, including building blocks such as checklists, technique descriptions, procedures, test organization structures, test environments, and test tools. It includes a complete toolbox to perform the testing, and its method is flexible and adaptive to different environments. There are four pillars of TMap:

- Test management (the test manager manages the test process)
- Complete toolbox (toolbox to perform the methods)
- Structured test process (the testing must go through lifecycle phases such as planning, preparation, specification, execution, and completion)
- Adaptive test method (TMap is flexible and it may be adapted to different environments including the Agile world).

The starting point for TMap is the business case, which provides the justification for the project. The testing process is then aligned to the business case, with the business characteristics translated into the process. The total test effort is related to the risks to the system, and the testing is focused on the parts of the system that are most important to the organization. The client is provided with adequate insight and control over the test process.

The test manager generally writes the master test plan (MTP), which defines the business-driven approach to the testing. The scope of the testing to be performed is agreed with the client, and the plan documents the scope of the testing and the deliverables to be produced. Test reports are produced during the project to highlight progress with the testing and the quality of the product.

TMap provides a flexible approach that can be adapted to many test situations and development methods. It is an adaptive method and is able to respond to a changing environment and respond to changing situations. It offers the tester a suite of practices such as test design techniques, test infrastructure, test strategy, test organization, and test tools, and the tester chooses the TMap elements that are appropriate for the testing.

The activities in the TMap lifecycle model are divided across seven phases where each phase consists of several activities. These are planning, control, setting up and maintaining infrastructure, preparation, specification, execution, and completion (Fig. 11.8). It is impossible to test the system completely, and so a risk-based approach to the test strategy and planning is adopted. The setting up and maintaining infrastructure is concerned with managing the test environments, test tools, and workplaces.

The specification phase is concerned with the specification of the required test cases, and the execution phase is concerned with the execution of the agreed tests to gain insight into the quality of the software. The completion phase is concerned with the preservation of testware such as the test cases and test environments for potential reuse, as well as the evaluation of the test process.

TMap has a complete toolbox to support the execution of the test process, which gives the tester a wider range of options for testing. The toolbox supports various techniques on how to test, as well as the infrastructure to support the testing. TMap considers the where, with, and what is to be tested and the organization that will do the testing.

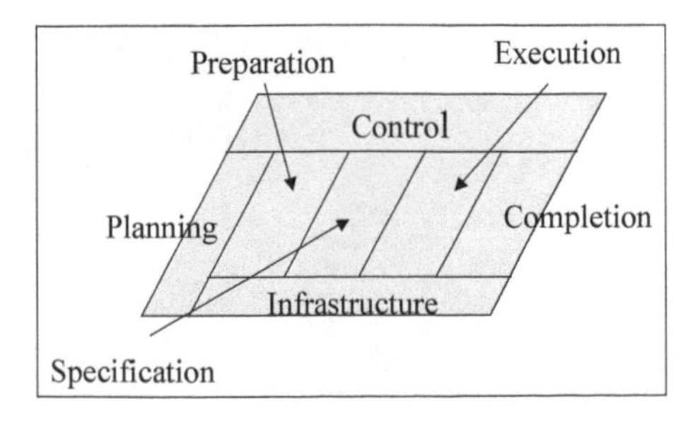


**Fig. 11.8** TMap lifecycle model

The test techniques include estimation, defect management, metrics, risk analysis, test design, and evaluation techniques. The infrastructure includes the test environments, tools, and workplaces that are needed to conduct the testing. The organization of the test team is concerned with the tasks and responsibilities of the test team and it must be an integral part of the project organization. There is more information on TMap in Tmaps (2004).

### 11.3.3 TPI Next Model

The *Test Process Improvement* (TPI) Model was developed by Sogeti (a subsidiary of Capgemini) in the late 1990s. It is a flexible and adaptive model designed to give insight into the maturity of the test process, and it is a business-driven approach to improving the process. TPI provides a framework to assess the current maturity of the testing process, and it identifies the strong and weak areas of the process. It provides an improvement path to improve the testing practices, and the model consists of:

- A maturity model
- Test maturity matrix
- Checklist
- Improvement suggestions.

The TPI model was revised to TPI Next in 2009, and it has 16 key areas including test strategy, test organization, test process management, estimation, reporting, defect management, test case design, test tools, and test environment (Fig. 11.9). Each key area can be at one of four levels of maturity and these are:

- Initial level (ad hoc)
- Controlled (doing the right things)
- Efficient (doing things the right way)
- Optimizing (continuously adapting).

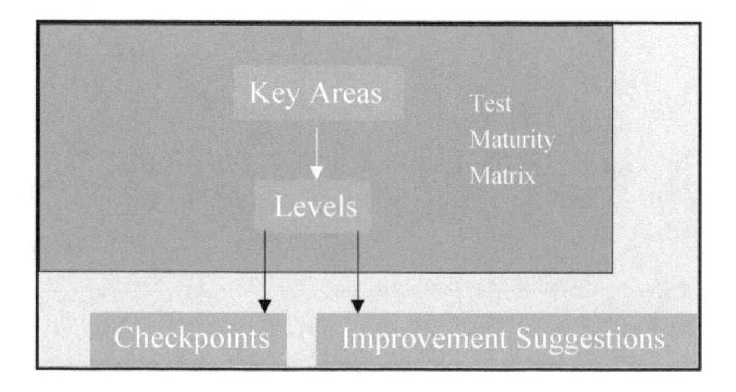


**Fig. 11.9** TPI model

Each maturity level for a key area includes a statement of what is expected at that level. It includes 2–4 checkpoints (or requirements) at that level and if the testing process passes all checkpoints, it is classified (assessed) to be performing at that level. Each maturity level includes a list of improvement suggestions to enable the checkpoints to be met, as well as 1–2 enablers that show how the test maturity of a key area can benefit from other disciplines by sharing best practice.

The test maturity matrix is a visual tool with the key areas running vertically from top to bottom and the four maturity levels running horizontally from left to right, and the checkpoint numbers fill the matrix cells. Once the particular checkpoint has been achieved the cell is shaded, and so the maturity matrix gives a visual picture of the current maturity of the key areas in the testing process.

After determining the levels for each key area attention is then devoted on which improvement steps to take, as the key areas and levels are not all equally important. The TPI Next model is adaptive to different lifecycle models including iterative development and Agile.

### 11.3.4 STEP Model

The *Systematic Test and Evaluation Process* (STEP) was developed by Bill Hetzel and David Gelperin in the mid-1980s as a way to implement the IEEE 829 standard for software test documentation. STEP is focused on software evaluation, which is a subdiscipline of software engineering concerned with ensuring that the software product does what it is supposed to do.

The main techniques employed in software evaluation are analysis, review, and testing, and the main focus of STEP is software testing with an emphasis on defect prevention.

STEP provides a process model for software testing which consists of tasks, work products, and roles and responsibilities. The model may be tailored or extended to meet the particular test situation, and there are three main phases of

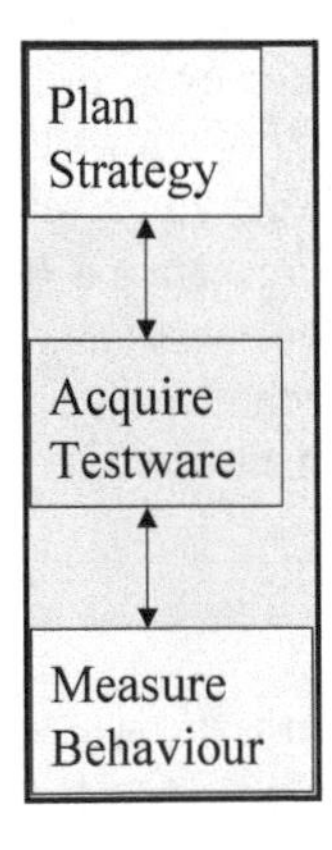


**Fig. 11.10** Phases of STEP model

STEP including planning the strategy for the testing, acquiring testware, and measuring behaviour (Fig. 11.10).

The planning phase is concerned with preparing a master test plan and detailed test plans. Acquiring testware is concerned with determining the test objectives, designing tests to evaluate the software, and implementing the plans and design. Finally, the measuring behaviour phase is concerned with executing the tests, checking the adequacy of the set of tests, and evaluating the software and testing process.

The STEP methodology does not assume any particular test organization or staffing. It specifies when the test activities and tasks should be performed, and the sequence in which the tasks are performed as well as what tasks to perform (e.g., plans and objectives come first; then, design and implementation; and finally, execution and evaluation).

The STEP methodology results in several test documents being produced (e.g., test plans, test design, test case specification, test log, and test report documents). There are four roles defined in the model (test manager, analyst, technician, and reviewer).

### 11.3.5 CTP Model

The *Critical Test Process* (CTP) model is a lightweight framework for test process improvement, which was developed by Rex Black of RBCS[4] in the early 2000s. It focuses on the test manager, the test team, and a few key test areas that must be done properly, as its philosophy is that some processes are critical whereas others are not. It seeks to focus on improvements based on their business value and avoids the approach of the large prescriptive models such as TMM or TPI. That is, it is concerned with the problem that the improvement solves, rather than implementing all of the practices of a particular maturity model. For example, the implementation of TMM may result in improvements to many areas that do not have problems, as well as to the areas that are problematic.

CTP is a non-prescriptive process model that describes the most important processes, and it identifies quantitative and qualitative attributes of good processes. However, it does not prescribe a particular order of improvement, and this flexibility allows the organization to focus on its specific challenges. CTP may then be used to select the order of improvements and to target the key areas that the organisation (test department) wishes to improve, based upon its needs and the business value of the improvements to the software testing process. Further, CTP is adaptive to many software development lifecycle models such as the Agile methodology.

[4]RBCS stands for Rex Black's Consulting Service, Inc. and it is a software testing and quality assurance company that helps organizations to improve their software testing.

CTP defines a critical test process as a process that is used frequently and affects the efficiency of the test team, and it is linked to project success in that it plays a key role in detecting defects and building confidence in the correctness of the software. There are 12 CTP critical processes and these include processes for test planning, test estimation, risk analysis, test team development, test execution, test release planning, bug reporting, result reporting, and change management.

CTP begins with an assessment of the existing test processes, which identifies which of the critical processes are done correctly and which need improvement. Various metrics may be examined during a CTP appraisal such as the defect detection percentage, ROI on testing, and requirements coverage. The assessor completes the assessment report and includes specific recommendations for improvement as well as the order in which they should be implemented. Improvement plans will then be made to implement the improvements.

### 11.3.6 PDCA Model

PDCA (plan, do check, act) is a systematic approach to problem-solving and process control (Fig. 1.3). It consists of four steps that are used for continuous process improvement, and these are plan, do, check, and act. It is also known as the *"PDCA Model"* or Shewhart's model. It is described in more detail in Table 1.2.

### 11.3.7 CMMI Model

The CMMI framework allows organizations to improve their maturity by improvements to their underlying processes (Chrissis 2011). It provides a clearly defined roadmap for improvement, and it allows the organization to improve at its own pace. Its approach is evolutionary rather than revolutionary, and it recognizes that a balance is required between project needs and process improvement needs. It allows the processes to evolve from ad hoc immature activities to disciplined mature processes.

The CMMI includes two process areas that are directly related to software testing. These are the verification and validation process areas which are concerned with ensuring that the system satisfies the requirements, is fit for purpose, and is what the user wants. Verification is concerned with ensuring that the work products reflect the specified requirements, whereas validation is concerned that the product will fulfil its intended use. That is, verification ensures that "You build it right", whereas validation ensures that "You built the right thing". The implementation of these process areas involves the implementation of processes for peer reviews and testing.

The purpose of the verification process area is to ensure that selected work products meet their specified requirements. The activities in this process area include preparation for verification; performing verification; identifying corrective

**Table 11.3** CMMI requirements for verification

| Specific goal | Specific practice | Description of specific practice/goal |
|---|---|---|
| SG 1 | | **Prepare for verification** |
| | SP 1.1 | Select products for verification |
| | SP 1.2 | Establish the verification environment |
| | SP 1.3 | Establish verification procedures and criteria |
| SG 2 | | **Perform peer reviews** |
| | SP 2.1 | Prepare for peer reviews |
| | SP 2.2 | Conduct peer reviews |
| | SP 2.3 | Analyse peer review data |
| SG 3 | | **Verify selected work products** |
| | SP 3.1 | Perform verification |
| | SP 3.2 | Analyse verification results |

**Table 11.4** CMMI requirements for validation

| Specific goal | Specific practice | Description of specific practice/goal |
|---|---|---|
| SG 1 | | **Prepare for validation** |
| | SP 1.1 | Select products for validation. |
| | SP 1.2 | Establish the validation environment |
| | SP 1.3 | Establish validation procedures and criteria |
| SG 2 | | **Ensure interface compatibility** |
| | SP 2.1 | Perform validation |
| | SP 2.2 | Analyse validation results |

action; and implementing the actions. Verification helps to ensure that the product will meet the customer and product requirements and is fit for purpose.

Peer reviews play a key role in verification and assist in identifying defects early in the software development lifecycle. They involve a rigorous examination of the work products by peers of the author, and the goal is to find defects as early as possible[5]. There are several types of peer reviews that may be carried out including software inspections and structured walkthroughs (see Chap. 4). The specific goals and practices of the verification process area are shown in Table 11.3:

The purpose of the validation process area is to demonstrate that a product or product components fulfils its intended use. The specific goals and practices for this process area are (Table 11.4).

Validation activities are applied to work products as well as products or product components and will demonstrate that the product or work product fulfils its

[5]There is a strong economic case for finding defects as early as possible, as the cost of correction increases the later the defect is found.

intended use. The validation activities are similar to the verification process and include analysis, testing, and inspections.

The validation environment should resemble the intended operating environment, and the validation activities is most effective when there is participation from the end-users as they have a deep understanding of how the system should perform. The key objective is to ensure that the right system is being built, as distinct from the verification activities, which confirm that the system is being built right.

## 11.4 Review Questions

1. What is a software process?
2. What is software process improvement?
3. What are the benefits of software process improvement?
4. What is test process improvement?
5. Describe the various models available for test process improvement?
6. Explain the TMM and TPI models?
7. What are the main barriers to successful software process improvement initiatives and how can they be overcome?
8. Describe the three phases in a SCAMPI appraisal.

## 11.5 Summary

Software process improvement helps software companies to deliver software on time and on budget. It plays a key role in helping companies to improve their software engineering capability, and it enables best practice in software engineering to be implemented. It has become an indispensable tool for software engineers and managers to achieve their goals, and it provides a return on investment.

Software process improvement initiatives lead to a focus on the process, which is important since many problems are caused by defective processes rather than by people. This allows companies to focus on fire prevention rather than firefighting, by critically examining their processes to determine whether they are fit for purpose. This leads to a culture of openness in discussing problems and instils process ownership among the process practitioners.

Test process improvement is concerned with the continuous improvement of the testing process. It offers a way to assess the maturity of the testing practices in an organization, and it provides a roadmap to improvement. This allows the organization to plan and improve the testing process to meet its needs.

There are several models to support test process improvement including TMM, TMap, TPI, STEP, CTP, PDCA and the CMMI. These models contain best practice in software testing that may be used to mature the test process.

## References

Chrissis, MB, Conrad M, Shrum S (2011) CMMI for development. Guidelines for process integration and product improvement, 3rd Edn. SEI series in software engineering. Addison Wesley, Boston

Tmaps (2004) TMap home pages. Sogeti Nederland B.V., Vianen. http://www.tmap.net

# Testing in the Agile World 12

**Key Topics**

Sprints
Stand-up meeting
Scrum
Stories
Refactoring
Pair programming
Software testing
Test-driven development
Continuous integration

## 12.1 Introduction

Agile is a popular lightweight software development methodology that aims to develop high-quality software faster than conventional approaches such as the waterfall development process (Fig. 12.1). Despite the fact that it is a lightweight methodology it does not mean that anything goes, and it is, in fact, a disciplined approach to software development. It emphasizes the following features:

- A collaborative style of working
- Integrated teams


G. O'Regan, *Concise Guide to Software Testing*,
Undergraduate Topics in Computer Science,
https://doi.org/10.1007/978-3-030-28494-7_12

**Fig. 12.1** Agile Dog. Creative Commons

- Frequent deliveries
- Ability to adapt to changing business needs.

Agile provides opportunities to assess the direction of a project throughout the development lifecycle. There has been a growth in interest in lightweight software development methodologies since the 1990 s, and these include approaches such as rapid application development (RAD), dynamic systems development method (DSDM), and extreme programming (XP). These approaches are referred to collectively as Agile methods.

Every aspect of Agile development such as requirements and design is continuously revisited during the development, and the direction of the project is regularly evaluated. Agile focuses on rapid and frequent delivery of partial solutions developed in an iterative and incremental manner. Each partial solution is evaluated by the product owner, and the feedback provided to determine the next steps for the project. Agile is more responsive to customer needs than traditional methods such as the waterfall model and its adherents believe that it results in:

- Higher quality
- Higher productivity
- Faster time to market
- Improved customer satisfaction.

It advocates adaptive planning, evolutionary development, early development, continuous improvement, and a rapid response to change. The term "*agile*" was coined by Kent Beck and others in the Agile Manifesto in 2001 (Beck 2001). The traditional waterfall model is similar to a wide and slow-moving value stream, and halfway through the project 100% of the requirements are typically 50% done. However, 50% of the requirements are typically 100% done halfway through an Agile project.

Agile has a strong collaborative style of working, and ongoing changes to requirements are considered normal in the Agile world. It argues that it is more realistic to change requirements regularly throughout the project, rather than attempting to define all of the requirements at the start of the project (as in the waterfall methodology). Agile includes controls to manage changes to the requirements, and good communication and early regular feedback is an essential part of the process.

A user story may be a new feature or a modification to an existing feature. The feature is reduced to the minimum scope that can deliver business value, and a feature may give rise to several stories. Stories often build upon other stories and the entire software development lifecycle is employed for the implementation of each story. Stories are either done or not done (i.e., there is no such thing as 50% done), and the story is complete only when it passes its acceptance tests.

Scrum is an Agile method for managing iterative development, and it consists of an outline planning phase for the project, followed by a set of sprint cycles (where each cycle develops an increment). *Sprint planning* is performed before the start of the iteration, and stories are assigned to the iteration to fill the available time. Each Scrum sprint is of a fixed length (usually 2–4 weeks), and it develops an increment of the system.

The estimates for each story and their priority are determined, and the prioritized stories are assigned to the iteration. A short (usually 15 min) morning stand-up meeting is held daily during the iteration, and it is attended by the Scrum Master, the project manager[1], and the project team. It discusses the progress made the previous day, problem reporting and tracking, and the work planned for the day ahead. A separate meeting is held for issues that require more detailed discussion.

Once the iteration is complete, the latest product increment is demonstrated to a review audience including the product owner. This is to receive feedback and to identify new requirements. The team also conducts a retrospective meeting to identify what went well and what went poorly during the iteration, as part of continuous improvement for future iterations. The planning for the next sprint then commences.

The Scrum Master is a facilitator who arranges the daily meetings and ensures that the Scrum process is followed. The role involves removing roadblocks so that the team can achieve their goals and communicating with other stakeholders. Agile employs pair programming and a collaborative style of working with the philosophy that two heads are better than one. This allows multiple perspectives in decision-making which provides a broader understanding of the issues.

[1]Agile teams are self-organizing and small teams (team size <20 people) do not usually have a project manager role, and the Scrum Master performs some light project management tasks.

Software testing is very important in verifying that the software is fit for purpose, and Agile generally employs automated testing for unit, acceptance, performance, and integration testing. Agile employs *test-driven development* with tests written before the code. The developers write code to make a test pass with ideally developers only coding against failing tests. This approach forces the developer to write testable code, as well as ensuring that the requirements are testable. Tests are run frequently with the goal of catching programming errors early. They are generally run on a separate build server to ensure that all the dependencies are checked. Tests are rerun before making a release.

Refactoring is employed in Agile as a design and coding practice. The objective is to change how the software is written without changing what it does. Refactoring is a tool for evolutionary design where the design is regularly evaluated, and improvements are implemented as they are identified. It helps in improving the maintainability and readability of the code and in reducing complexity. The automated test suite is essential in demonstrating that the integrity of the software is maintained following refactoring.

Continuous integration allows the system to be built with every change. Early and regular integration allows early feedback to be provided, and it also allows all of the automated tests to be run thereby identifying problems earlier. The main philosophy and features of Agile are:

- Working software is more useful than documents
- Direct interaction is preferred over documentation
- Change is accepted as a normal part of life in the Agile world
- Customer is involved throughout the project
- Demonstrate value early
- Feedback and adaptation are employed in decision-making
- Aim is to achieve a narrow fast flowing value stream
- User Stories and sprints are employed
- A project is divided into iterations
- An iteration has a fixed length (i.e., Time boxing is employed)
- Entire software development lifecycle is employed for implementation of the story
- Stories are either done or not done (no such thing as 50% done)
- Iterative and incremental development is employed
- Emphasis on quality
- Stand-up meetings held daily
- Rapid conversion of requirements into working functionality
- Delivery is made as early as possible
- Maintenance is seen as part of the development process
- Refactoring and evolutionary design employed
- Continuous integration is employed
- Short cycle times
- Plan regularly
- Early decision-making.

Stories are prioritized based on a number of factors including:

- Business value of story
- Mitigation of risk
- Dependencies on other stories.

## 12.2 Scrum Methodology

Scrum is a framework for managing an Agile software development project (Fig. 12.2). It is not a prescriptive methodology as such, and it relies on a self-organizing, cross-functional team to take the feature from idea to implementation. The cross-functional team includes the *product owner* who represents the interest of the users and ensures that the right product is built; the *Scrum Master* who is the coach for the team and helps the team to understand the Scrum process and to perform at the highest level, as well as performing some light project management activities such as project tracking; the self-organizing *team* itself that decides on which person should work on which tasks; and so on.

The Scrum methodology breaks the software development for the project into a series of sprints, where each sprint is of fixed time duration of 2–4 weeks. There is a planning meeting at the start of the sprint where the team members determine the number of items/tasks that they can commit to and then create a sprint backlog (*to do list*) of the tasks to be performed during the sprint. The Scrum team takes a small set of features from idea to coded and tested functionality that is integrated into the evolving product.

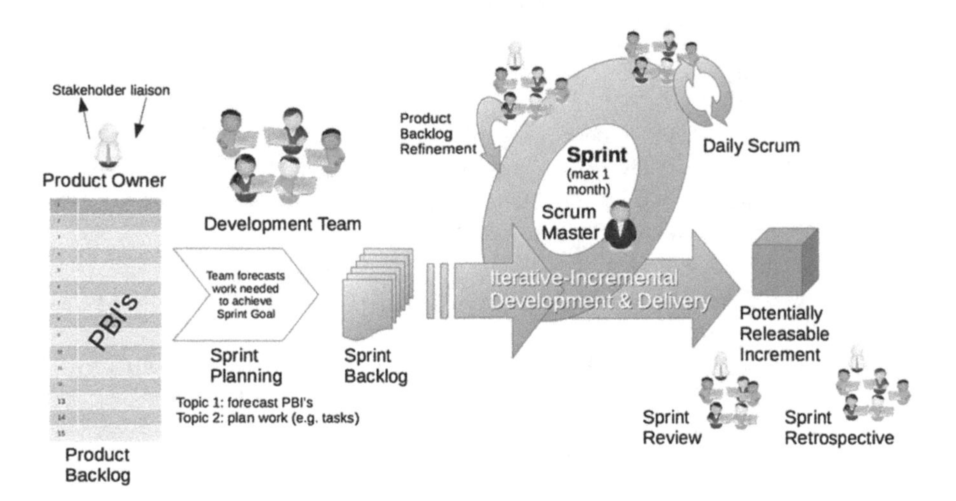


**Fig. 12.2** Scrum framework. Creative Commons

The team attends a daily stand-up meeting (usually for 15 min) where the progress of the previous day is discussed, as well as any obstacles to progress. The new functionality is demonstrated to the product owner and any other relevant stakeholders at the end of the sprint, and this may result in changes to the delivered functionality or the addition of new items to the product backlog. There is a sprint retrospective meeting to reflect on the sprint and to identify improvement opportunities.

The main deliverable produced using the Scrum framework is the *product itself*, and Scrum expects to build a properly tested product increment (in a shippable state) at the end of each sprint. The *product backlog* is another deliverable and it is maintained and prioritized by the product owner. It is a complete list of the functionality (user stories) to be added to the product, and there is also the *sprint backlog* which is the list of functionality to be implemented in the sprint. Other deliverables are the *sprint burnout* and *release burnout* charts, which show the amount of work remaining in a sprint or release and indicate the extent to which the sprint or release is on schedule.

The Scrum Master is the expert on the Agile process and acts as a coach to the team thereby helping the team to achieve a high level of performance. The role differs from that of a project manager, as the Scrum Master does not assign tasks to individuals or provide day-to-day direction to the team. However, the Scrum Master typically performs some light project management tasks.

Many of the traditional project manager responsibilities such as task assignment and day-to-day project decisions revert back to the team, and the responsibility for the scope and schedule trade-off goes to the product owner. The product owner creates and communicates a solid vision of the product and shares the vision through the product backlog. Larger Agile projects (team size >20) will often have a dedicated project manager.

### 12.2.1 User Stories

A *user story* is a short simple description of a feature written from the viewpoint of the user of the system. They are often written on index cards or sticky notes and arranged on walls or tables to facilitate discussion. This approach facilitates the discussion of the functionality rather than the written text.

A user story can be written at varying levels of detail, and a large detailed user story is known as an epic. An epic story is often too large to be implemented in one sprint, and such a story is often split into several smaller user stories.

It is the product owner's responsibility to ensure that a product backlog of user stories exist, but the product owner is not required to write all stories. In fact, anyone can write a user story, and each team member usually writes a user story during an Agile project. User stories are written throughout an Agile project, with a user story-writing workshop generally held at the start of the project. This leads to the product backlog that describes the functionality to be added during the project. Some of these will be epics, and these will need to be decomposed into smaller

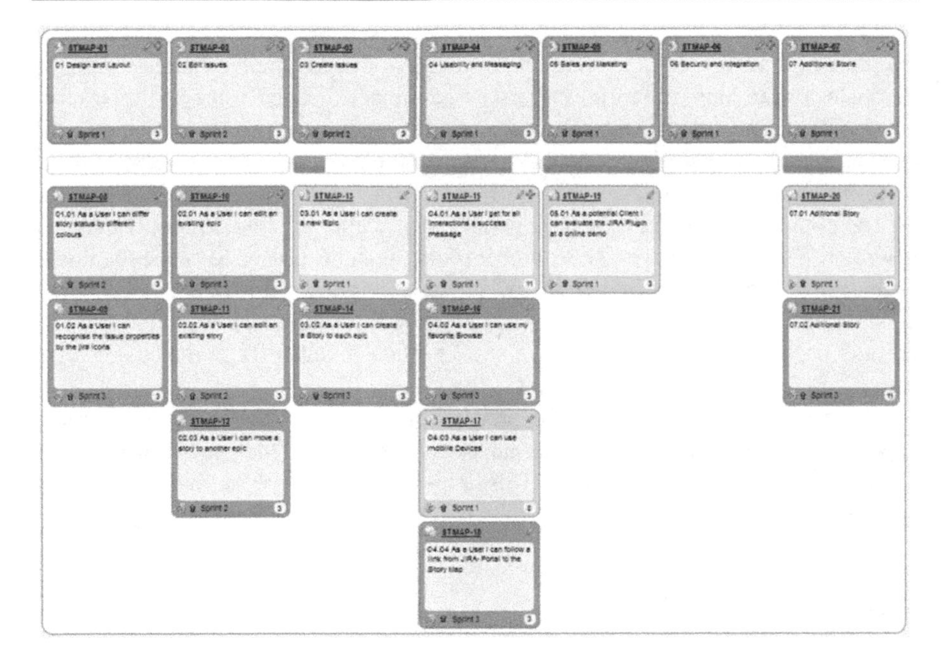

**Fig. 12.3** User story map

stories that will fit into the time-boxed sprint. New user stories may be written at any time and added to the product backlog (the user story map in Fig. 12.3 is a two-dimensional representation of the product backlog).

There is no requirements document as such in Agile, and the product backlog (i.e., the prioritized list of functionality of the product to be developed) is closest to the idea of a requirements document for a traditional project. However, the written part of a user story in Agile is incomplete until the discussion of that story takes place. It is often useful to think of the written part of a story as a pointer to the real requirement, such as a diagram showing a workflow or the formula for a calculation.

## 12.2.2 Estimation in Agile

Planning poker is a popular consensus-based estimation technique often used in Agile, and it is used to estimate the effort required to implement a user story. The planning session starts with the product owner reading the user story or describing a feature to the estimators.

Each estimator holds a deck of planning poker cards with values like 0, 1, 2, 3, 5, 8, 13, 20, 40, and 100, where the values represent the units in which the team estimates. The estimators discuss the feature with the product owner, and when the discussion is fully complete and all questions answered, each estimator privately selects a card to reflect his or her estimate.

All cards are then revealed and if all values are the same then that value is chosen as the estimate. Otherwise, the estimators discuss their estimates with the rationale for the highest and lowest discussed in detail. Each estimator then reselects an estimate card, and the process continues until consensus is achieved, or if consensus cannot be achieved the estimation of the particular item is deferred until more information is available.

The initial estimation session usually takes place after the initial product backlog is written. This session may take a number of days, and it is used to create the initial estimates of the size and scope of the project. Further estimation and planning sessions take place regularly during the project as user stories are added to the product backlog, and these will typically take place towards the end of the current sprint.

The advantage of the estimation process employed is that it brings multiple expert opinions from the cross-functional team together, and the experts justify their estimates in the detailed discussion. This helps to improve the estimation accuracy in the project.

### 12.2.3 Pair Programming

Pair programming is an Agile technique where two programmers work together at one computer (Fig. 12.4). The author of the code is termed the *driver*, and the other programmer is termed the *observer* (or *navigator*) and is responsible for reviewing each line of written code. The observer also considers the strategic direction of the coding and proposes improvement suggestions and potential problems that may

**Fig. 12.4** Pair programming. Creative Commons

need to be addressed. The driver can focus on the implementation of the current task and use the observer as a safety net. The two programmers switch roles regularly during the development of the new functionality.

Pair programming requires more programming effort to develop code compared to programmers working individually. However, the resulting code is generally of higher quality, with fewer defects and a reduction in the cost of maintenance. Further, pair programming enables a better design solution to be created as more design alternatives are considered.

This is since two programmers are bringing different experiences to the problem, and they may have different ways of solving the problem. This leads them to explore a larger number of ways of solving the problem than an individual programmer. Finally, pair programming is good for knowledge sharing and learning, including knowledge on programming practice and design and knowledge about the system among the team.

## 12.3 Software Testing in Agile

Software testing is essential in verifying that the software is fit for purpose and that it is ready to be released to the customer. Conventional software projects employ a testing phase to verify the correctness of the software, and the testing also ensures that the defects identified during testing have been resolved. The developers and testers are in a sense in different silos during a conventional project, which potentially leads to an adversarial relationship between them. However, in the Agile world, testing is employed from the very beginning of the project to provide regular feedback on the extent to which the product meets business needs. The developers and testers are very much part of the one integrated team, and they work closely together in a spirit of collaboration. That is, *there is a completely different mindset to testing employed in the Agile world.*

Testing is the responsibility of the test group in a conventional project, whereas testing is the responsibility of the entire team in an Agile project. Conventional projects employ a testing phase where various types of testing such as system and performance testing may be performed. However, Agile projects employ continuous testing from the start of the project, and this helps in ensuring that continuous progress is made during the sprint, and that the features have been correctly implemented.

It is fundamental in Agile that all the features be completely tested (including UAT testing) during the sprint (i.e., all the testing must be done), *as any features that have not been completely tested are considered to be not done.* This may result in the team being unable to do as much in the sprint as previously thought, as the team is not going as fast as they thought and everyone tests to eliminate the bottleneck.

For conventional projects, there is often a large gap in time between development and testing of the software, and this increases the risk to the quality of the project. However, in the Agile world, teams test early and test often, which provides a short feedback loop on the quality of the software. That is, the team knows early whether there are problems with the software, whereas conventional projects learn about problems very late in the project. Agile's approach is to keep the code clean and genuine defects are corrected as they are identified.

Agile projects are ready to test early in the sprint, and automated tests (including unit and regression) are run frequently to provide rapid feedback. Automated regression tests are very useful in finding problems quickly, and this helps to reduce risk and rework. Manual testing (e.g., exploratory or regression) takes longer to execute and may require one or more team members being available for several days.

Conventional projects produce a suite of comprehensive test documentation including test plans, test case specifications, test reports, and so on. However, in the Agile world, lightweight test documentation is employed with Agile testers using reusable checklists to suggest tests and using lightweight documentation tools.

Agile generally employs automated testing for unit, acceptance, performance, and integration testing. Conventional projects employ a "*test-last*" approach with the requirements and design coming first and the tests derived from them and the testing taking place at the end of the project. Agile employs a "*test-first*" approach with the tests defined with the requirements and used to drive the development effort.

That is, Agile employs *test-driven development* with tests written before the code. The developers write code to make a test pass with ideally developers only coding against failing tests. The code may then be refactored to improve its maintainability and readability and retested. Test-driven development forces the developer to write testable code, as well as ensuring that the requirements are testable. Tests are run frequently with the goal of catching programming errors early, and they may be run on a separate build server to ensure that all dependencies are checking prior to making a release.

Agile employs automated unit/integration tests which are written by the programmer and are executed frequently and especially following change. It employs automated system tests that define the externally expected behaviour of the system and these tests are executed regularly as part of continuous integration. Exploratory testing is employed as an Agile practice to learn about the software by designing and executing tests and may be used to target vulnerabilities in the system.

### 12.3.1 Test-Driven Development

Test-driven development (TDD) was developed by Kent Beck and others as part of extreme programming, and the developers focus on testing the requirements before writing the code. The application is written with testability in mind, and the developers must consider how to test the application in advance. Further, it ensures that test cases for every feature are written, and writing tests early help in gaining a deeper understanding of the requirements.

TDD is based on the transition of the requirements into a set of test cases, and the software is then written to pass the test cases. In other words, the test-driven development of a new feature begins with writing a suite of test cases based on the requirements for the feature, and the code for the feature is written to pass the test cases. *This is a paradigm shift from traditional software engineering* where the unit tests are written and executed after the code is written.

The tests are written for the new feature, and initially, all tests fail as no code has been written, and so the first step is to write some code that enables the new test cases to pass. This new code may be imperfect (it will be improved later), but this is acceptable at this time as the only purpose is to pass the new test cases. The next step is to ensure that the new feature works with the existing features, and this involves executing all new and existing test cases.

This may involve modification of the source code to enable all of the tests to pass and to ensure that all features work correctly together. The final step is refactoring the code, and this involves cleaning up and restructuring the code, and improving its structure and readability. The test cases are rerun during the refactoring to ensure that the functionality is not altered in any way. The process repeats with the addition of each new feature.

Continuous integration allows the system to be built with every change, and this allows early feedback to be provided. It also allows all of the automated tests to be run, thereby ensuring that the new feature works with the existing functionality, and identifying problems earlier.

### 12.3.2 Agile Test Principles

There are several test principles employed in the Agile methodology including that it is a test-driven approach with testing continuous rather than sequential as in the waterfall model, and it is performed by the integrated team rather than a dedicated test team. Continuous testing shortens the time for feedback to be provided, and the code is kept clean and simplified since all defects are corrected within the sprint. Agile uses lightweight documentation for testing (reusable checklists) in order to focus on the tests. The test principles are summarized in Table 12.1:

**Table 12.1** Agile test principles

| Principle | Description |
|---|---|
| Testing provides feedback | Testing is used to provide feedback and visibility to move the project forward |
| Continuous testing | Testing is a way of life in Agile and it takes place frequently during the sprint |
| Testing by entire team | Both developers and testers execute tests with the whole team becoming involved to eliminate bottlenecks in testing |
| Short feedback loop | Agile teams test early and test often to obtain rapid feedback on how the software is behaving |
| Clean code | Developers fix genuine defects as they are found thereby keeping the code clean |
| Lightweight documentation | Testers use reusable checklists and lightweight documentation tools |
| Done means "done" | A feature is not complete until it has been fully implemented and tested |
| Test-driven | The tests are defined with the requirements and used to drive the development efforts |

## 12.4 Review Questions

1. What is Agile?
2. How does Agile differ from the traditional waterfall model?
3. What is a user story?
4. Explain how estimation is done in Agile.
5. What is test-driven development?
6. Describe the Scrum methodology and the role of the Scrum Master.
7. Explain how testing is performed in the Agile world.
8. Explain pair programming and describe its advantages.
9. What are the strengths and weaknesses of the Agile methodology?
10. Explain the principles of testing in the Agile world.

## 12.5 Summary

This chapter gave a brief introduction to Agile and to testing in the Agile world. Agile is a popular lightweight software development methodology that advocates adaptive planning, evolutionary development, early development, continuous improvement, and a rapid response to change. The traditional waterfall model is similar to a wide and slow-moving value stream, and halfway through the project

100% of the requirements are typically 50% done. However, 50% of the requirements are typically 100% done halfway through an Agile project.

Agile has a strong collaborative style of working, and ongoing changes to requirements are considered normal in the Agile world. It includes controls to manage changes to the requirements, and good communication and early regular feedback is an essential part of the process.

A story may be a new feature or a modification to an existing feature. It is reduced to the minimum scope that can deliver business value, and a feature may give rise to several stories. Stories often build upon other stories and the entire software development lifecycle is employed for the implementation of each story. Stories are either done or not done and the story is complete only when it passes its acceptance tests.

The Scrum approach is an Agile method for managing iterative development, and it consists of an outline planning phase for the project followed by a set of sprint cycles (where each cycle develops an increment). Each Scrum sprint is of a fixed length (usually 2-4 weeks), and it develops an increment of the system.

The estimates for each story and their priority are determined, and the prioritized stories are assigned to the iteration. A short (usually 15 min) morning stand-up meeting is held daily during the iteration and attended by the project manager and the project team. It discusses the progress made the previous day, problem reporting and tracking, and the work planned for the day ahead.

Software testing is employed from the very beginning of an Agile project to provide regular feedback on the extent to which the product meets business needs. The developers and testers are part of one integrated team, and they work closely together in a spirit of collaboration. There is a completely different mindset to testing employed in the Agile world.

Once the iteration is complete, the latest product increment is demonstrated to a review audience including the product owner. This is to receive feedback and to identify new requirements. The team also conducts a retrospective meeting to identify what went well and what went poorly during the iteration, as part of continuous improvement for future sprints.

## Reference

Beck K, et al (2001) Manifesto for Agile Software Development. Agile Alliance. http://agilemanifesto.org/

# Verification of Safety-Critical Systems 13

**Key Topics**

Software Reliability
Dependability
Safety-Critical Systems
Cleanroom
Vienna Development Method
Z Specification Language
Model-Oriented Approach
Axiomatic Approach
Refinement

## 13.1 Introduction

The release of an unreliable software product may result in damage to property or injury (including loss of life) to a third party. Consequently, companies need to be confident that their software products are fit for purpose prior to their release. It is essential that software that is widely used is dependable, which means that the software is available whenever required, and that it operates safely and reliably without any adverse side effects.

Today, billions of devices and computers are connected to the Internet and this has led to a growth in attacks on computers. It is essential that computer security is carefully considered, and that developers are aware of the threats facing a system, and techniques to eliminate them. The software developers need to be able to


G. O'Regan, *Concise Guide to Software Testing*,
Undergraduate Topics in Computer Science,
https://doi.org/10.1007/978-3-030-28494-7_13

develop secure dependable systems that are able to deal with and recover from external attacks.

A safety-critical system is a system whose failure could result in significant economic damage or loss of life. There are many examples of safety-critical systems such as aircraft flight control systems, nuclear power stations, and missile systems. It is essential to employ rigorous processes in their design and development, and software testing alone is usually insufficient in verifying the correctness of such systems (Fig. 13.1).

The safety-critical industry takes the view that any change to safety-critical software creates a new program. The new program is therefore required to demonstrate that it is reliable and safe to the public, and so extensive testing needs to be performed. Other techniques such as formal verification and model checking may be employed to provide an extra level of assurance in the correctness of the system.

Safety-critical systems need to be reliable, dependable, and available for use whenever required. The software must operate correctly and reliably without any adverse side effects. The consequence of failure (e.g. the failure of a weapons system) could be massive damage, leading to loss of life or endangering the lives of the public.

The development of a safety-critical system needs to be rigorous, and subject to strict quality assurance to ensure that the system is safe to use and that the public will not be in danger. This involves rigorous design and development processes to minimize the number of defects in the software, as well as comprehensive testing to verify its correctness. It may not always be possible to test the safety-critical system under real-world conditions, and in such situations, it is common to employ other

**Fig. 13.1** Grafenrheinfeld Nuclear Power Plant. Germany. Creative Commons

techniques to provide increased confidence in its correctness. Formal methods are one approach that assists in the development and verification of safety-critical systems.

Formal methods consist of a set of mathematical techniques to rigorously state the requirements of the proposed system. They may be employed to derive a program from its mathematical specification and to provide a rigorous proof that the implemented program satisfies its specification. They provide the facility to prove that certain properties are true of the specification, and this is valuable, especially for safety-critical and security-critical applications. A mathematical specification is not subject to the ambiguities inherent in a natural language description of a system, and it may be subjected to a rigorous analysis to demonstrate the presence or absence of key properties.

Safety-critical systems are generally designed for fault tolerance, where the system can deal with (and recover from) faults that occur during execution. Fault tolerance is achieved by anticipating exceptional events, and in designing the system to handle them. A fault-tolerant system is designed to fail safely, and programs are designed to continue working (possibly at a reduced level of performance) rather than crashing after the occurrence of an error or exception. Many fault-tolerant systems mirror all operations, where each operation is performed on two or more duplicate systems, and so if one fails then the other system can take over.

## 13.2 Software Reliability

Software reliability is the probability that the program works without failure for a period of time, and it is usually expressed as the mean time to failure. It is different from hardware reliability, in that hardware is characterized by components that physically wear out, whereas software is intangible and software failures are due to design and implementation errors. In other words, software is either correct or incorrect when it is designed and developed, and it does not physically deteriorate over time.

The hardware field has been very successful in developing sound reliability models, which allows useful predictions of how long a hardware component (or product) will function. This has led to a growing interest in the software field in the development of a scientific software reliability model. Such a model would provide a sound mechanism to predict the reliability of the software prior to its deployment at the customer site, as well as providing confidence that the software is fit for purpose and safe to use.

**Definition 13.1** (*Software Reliability*) *Software reliability* is the probability that the program works without failure for a specified length of time, and it is a statement of the future behaviour of the software. It is generally expressed in terms of the *mean-time-to-failure* (MTTF) or the *mean-time-between-failure* (MTBF).

Statistical sampling techniques are often employed to predict the reliability of hardware, as it is not feasible to test all items in a production environment. The quality of the sample is used to make inferences on the quality of the entire population, and this approach is effective in manufacturing environments where variations in the manufacturing process often lead to defects in the physical products.

A hardware failure generally arises due to a component wearing out and often a replacement component is required. Hardware components are expected to last for a certain period of time, and the variation in the failure rate of a hardware component is often due to variations in the manufacturing process, or to the operating environment of the component. Good hardware reliability predictors have been developed, and each hardware component has an expected mean time to failure. The reliability of a product may be determined from the reliability of the individual components of the hardware.

Software is an intellectual undertaking involving a team of designers and programmers. It does not physically wear out as such, and software failures manifest themselves from particular user inputs. Each copy of the software code is identical, and the software code is either correct or incorrect. That is, software failures are due to design and implementation errors rather than to the software physically wearing out over time. A number of software reliability models (e.g. the software reliability growth models) have been developed, but the software engineering community has not yet developed a sound software reliability predictor model that can be trusted.

The software population to be sampled consists of all possible execution paths of the software, and since this is potentially infinite it is generally not possible to perform exhaustive testing. The way in which the software is used (i.e. the inputs entered by the users) will impact upon its perceived reliability. Let $I_f$ represent the fault set of inputs (i.e. $i_f \in I_f$ if and only if the input of $i_f$ by the user leads to failure). The randomness of the time to software failure is due to the unpredictability in the selection of an input $i_f \in I_f$. It may be that the elements in $I_f$ are inputs that are rarely used and that the software will be perceived as being reliable.

Harlan Mills and others showed that *coverage testing* is not as cost effective as *usage testing* in increasing MTTF (Cobb and Mills 1990). Statistical usage testing may be used to make predictions on the future performance and reliability of the software. It requires an understanding of the expected usage profile of the system and the population of all possible usages of the software. The sampling is done in accordance with the expected usage profile, and a software reliability measure is calculated.

Harlan Mills and others at IBM developed the Cleanroom approach to software development (O'Regan 2006). This formal approach to software development involves the application of statistical techniques to calculate a software reliability measure of the software based on its expected use.[1] This involves executing tests chosen from the population of all possible uses of the software in accordance with

[1]The expected usage of the software (or operational profile) is a quantitative characterization (usually based on probability) of how the system will be used.

**Table 13.1** Software reliability testing

| Item | Formula | Description |
|---|---|---|
| Availability | $\text{Availability} = \frac{\text{MTBF}}{\text{MTBF} + \text{MTTR}}$ | It is the percentage of the time that the software system is running |
| Mean time between failure | $\text{MTBF} = \frac{\text{Sample Interval Time}}{\#\text{Outages}}$ | Average length of time between outages |
| Mean time to repair | $\text{MTTR} = \frac{\text{Total Outage Time}}{\#\text{Outages}}$ | Average length of time that it takes to correct the outage (average duration of outage) |

the probability of its expected use. Statistical usage testing is more effective than coverage testing in finding defects that lead to failure.

Software reliability models are an attempt to predict the future reliability of the software and in deciding on whether the software is ready for release. A defect does not always result in a failure, as it may occur on a rarely used execution path. Studies indicate that many observed failures arise from a small proportion of the existing defects.

The defect count and defect density may be poor predictors of operational reliability, and an emphasis on removing a large number of defects from the software may not be sufficient to achieve high reliability. The correction of defects in the software leads to a newer version of the software, and reliability models assume reliability growth, i.e. the new version is more reliable than the older version as several identified defects have been corrected. The safety-critical industry (e.g. the nuclear power industry) takes the conservative viewpoint that any change to a program creates a new program. The new program is therefore required to demonstrate its reliability, and so extensive testing needs to be performed before any conclusions may be drawn.

There is a need to be careful with *reliability growth models*, as there is no tangible growth in reliability unless the corrected defects are likely to manifest themselves as a failure.[2] Many existing software reliability growth models assume that all remaining defects in the software have an equal probability of failure and that the correction of a defect leads to an increase in software reliability. These assumptions are questionable.

Software reliability testing is concerned with testing to determine the extent to which the software functions correctly for a given period of time (Table 13.1). Software reliability is the probability that the software works correctly for a given period of time, and it is calculated from the failure rate $\lambda$ = 1/MTTF[3] and the reliability function $R(t) = e^{-\lambda t}$.

[2]We are assuming that the defect has been corrected perfectly with no new defects introduced by the changes made.

[3]MTBF = MTTF + MTTR.

## 13.3 Software Dependability

It is essential that software that is widely used is dependable (or trustworthy). In other words, the software should be available whenever required, as well as operating properly, safely, and reliably, without any adverse side effects or security concerns. This is especially true of the software used in the safety-critical and security-critical fields, as the consequence of failure (e.g. the failure of a nuclear power plant) could be catastrophic leading to massive damage leading or loss of life.

Dependability engineering is concerned with techniques to improve the dependability of systems, and it involves the use of a rigorous design and development process to minimize the number of defects in the software. A dependable system is generally designed for fault tolerance, where the system can deal with (and recover from) faults that occur during software execution. Such a system needs to be secure, and able to protect itself from accidental or deliberate external attacks. Table 13.2 lists a number of dimensions to dependability.

Modern software systems are subject to attack by malicious software such as viruses that may change its behaviour, or corrupt data making the system unreliable. Other malicious attacks include a denial of service attack that negatively impacts the system's availability.

The design and development of dependable software need to include protection measures to prevent against such external attacks that compromise the availability and security of the system. Further, a dependable system needs to include recovery mechanisms to enable normal service to be restored as quickly as possible following an attack.

Dependability engineering is concerned with techniques to improve the dependability of systems and in designing dependable systems. A dependable system will generally be developed using an explicitly defined repeatable process, and it may employ redundancy (spare capacity) and diversity (different types) to achieve reliability.

There is a trade-off between dependability and system performance, as dependable systems will need to carry out extra checks to monitor themselves and to check for erroneous states, and to recover from faults before failure occurs. This inevitably leads to increased costs in the design and development of dependable systems.

**Table 13.2** Dimensions of dependability

| Dimension | Description |
|---|---|
| Availability | The system is available for use at any time |
| Reliability | The system operates correctly and is trustworthy |
| Safety | The system operates safely and does not injure people or damage the environment |
| Security | The system is secure and prevents unauthorized intrusions |

Software availability is the percentage of the time that the software system is running and is a measure of the uptime/downtime of the software during a particular time period. The downtime refers to a period of time when the software is unavailable for use (including planned and unplanned outages), and many companies aim to develop software that is available for use 99.999% of the time in the year (i.e. an annual downtime of less than 5 min per annum). This goal is known as *five nines*, and it is a common goal in the telecommunications sector. We discussed availability metrics in Chap. 9.

Safety-critical systems are systems where it is essential that the system is safe for the public, and that people or the environment is not harmed in the event of system failure. The failure of a safety-critical system could in some situations lead to loss of life or serious economic damage.

Formal methods provide a precise way of specifying the requirements and demonstrating (using mathematics) that key properties are satisfied in the formal specification. They may be used to show that the implemented program satisfies its specification, and their use leads to increased confidence in the correctness of dependable systems.

The security of the system refers to its ability to protect itself from accidental or deliberate external attacks, which are common today since most computers are networked and connected to the Internet. There are various security threats in any networked system including threats to the confidentiality and integrity of the system and its data, and threats to the availability of the system.

Therefore, controls are required to enhance security and to ensure that attacks are unsuccessful. Encryption is one way to reduce system vulnerability, as encrypted data is unreadable to the attacker. There may be controls that detect and repel attacks, which are used to monitor the system and to take action to shut down parts of the system or restrict access in the event of an attack. There may be controls that limit exposure (e.g. insurance policies and automated backup strategies) that allow recovery from the problems introduced.

It is important to have a reasonable level of security as otherwise all of the other dimensions of dependability (reliability, availability, and safety) are compromised. Security loopholes may be introduced in the development of the system, and so care needs to be taken to prevent hackers from exploiting security vulnerabilities.

Risk analysis plays a key role in the specification of security and dependability requirements, and this involves identifying risks that can result in serious incidents. This leads to the generation of specific security requirements as part of the system requirements to ensure that these risks do not materialize, or if they do materialize then serious incidents will not materialize.

**Fig. 13.2** Formal signing of the treaty of Versailles in 1919. Public Domain

## 13.4 Formal Methods

The term "*formal*" is used to refer to form, structure or rules rather than content, and examples include a formal dance or a formal meeting (Fig. 13.2). The term "*formal methods*" refer to various mathematical techniques used for the formal specification and development of software. They consist of a formal specification language and employ a collection of tools to support the syntax checking of the specification, as well as the proof of properties of the specification. They allow questions to be asked about what the system does independently of the implementation.

The use of mathematical notation avoids speculation about the meaning of phrases in an imprecisely worded natural language description of a system. Natural language is inherently ambiguous, whereas mathematics employs a precise rigorous notation. Spivey (1992) defines formal specification as:

**Definition 13.1** (*Formal Specification*) Formal specification is the use of mathematical notation to describe in a precise way the properties that an information system must have without unduly constraining the way in which these properties are achieved.

The formal specification thus becomes the key reference point for the different parties involved in the construction of the system. It may be used as the reference point for the requirements; program implementation; testing and program documentation. It promotes a common understanding for all those concerned with the system. The term "*formal methods*" is used to describe a formal specification language and a method for the design and implementation of a computer system. Formal methods may be employed at a number of levels:

- Formal specification only (program developed informally)
- Formal specification, refinement, and verification (some proofs)
- Formal specification, refinement, and verification (with extensive theorem proving).

The specification is written in a mathematical language, and the implementation may be derived from the specification via step-wise refinement.[4] The refinement step makes the specification more concrete and closer to the actual implementation. There is an associated proof obligation to demonstrate that the refinement is valid and that the concrete state preserves the properties of the abstract state. Thus, assuming that the original specification is correct and the proofs of correctness of each refinement step are valid, then there is a very high degree of confidence in the correctness of the implemented software.

Step-wise refinement is illustrated as follows: the initial specification S is the initial model $M_0$; it is then refined into the more concrete model $M_1$, and $M_1$ is then refined into $M_2$, and so on until the eventual implementation $M_n = E$ is produced.

$$S = M_0 \sqsubseteq M_1 \sqsubseteq M_2 \sqsubseteq M_3 \sqsubseteq \ldots\ldots \sqsubseteq M_n = E$$

Requirements are the foundation of the system and irrespective of the best design and development practices; the product will be incorrect if the requirements are incorrect. The objective of requirements validation is to ensure that the requirements reflect what is actually required by the customer (in order to build the right system). Formal methods may be employed to model the requirements, and the model exploration yields further desirable or undesirable properties.

Formal methods provide the facility to prove that certain properties are true of the specification, and this is valuable, especially in safety-critical and security-critical applications. The properties are a logical consequence of the mathematical requirements and the requirements may be amended where appropriate. Thus, formal methods may be employed in a sense to debug the requirements during requirements validation.

The use of formal methods generally leads to more robust software and to increased confidence in its correctness. They may be employed at different levels (e.g. it may just be used for specification with the program developed informally). The use of formal methods does not eliminate the need for software testing, but their use provides additional confidence in the correctness of the implemented system. The challenges involved in the deployment of formal methods in an organization

[4]It is questionable whether step-wise refinement is suitable in mainstream software engineering, as it involves re-writing a specification several times and takes significant time to prove that the refinement steps are valid. It is more relevant to the safety-critical field.

include the education of staff in formal specification, as the use of these mathematical techniques may be a culture shock to many staff.

Formal methods have been applied to several areas, especially the safety- and security-critical fields, to develop reliable and dependable software. The applications include the verification of software in the railway sector, microprocessor verification, the specification of standards, and the specification and verification of programs (Hinchey and Bowen 1995).

The use of a formal method such as Z or VDM forces the software engineer to be precise, and this helps in avoiding the ambiguities present in natural language (Bjorner and Jones 1982; Diller 1990). Clearly, a formal specification should be subject to peer review to provide confidence in its correctness. Formal methods are potentially quite useful and reasonably easy to use. However, new formalisms need to be intuitive to be usable, as some of the formalisms introduced have been a culture shock to users. There are advantages in using classical mathematics as the notation, since mathematical notation is intuitive and familiar to high-school students.

## 13.5 Cleanroom Methodology

Harlan Mills and others at IBM developed the Cleanroom methodology as a way to develop high-quality software (Cobb and Mills 1990). Cleanroom helps to ensure that the software is released only when it has achieved the desired quality level, and the probability of zero-defects is very high. The name "cleanroom" comes from specialized industrial production in the microprocessor and pharmaceutical sector (Fig. 13.3).

The way in which the software is used will impact on its perceived quality and reliability. Failures will manifest themselves on certain input sequences only, and as users often employ different input sequences, each user may have a different perception of the reliability of the software. The knowledge of how the software will be used allows the software testing to focus on verifying the correctness of common everyday tasks carried out by users.

This means that it is important to determine the operational profile of users to enable effective software testing to be performed. The operational profile may be difficult to determine and it could change over time, as users may change their behaviour as their needs evolve. The determination of the operational profile involves identifying the common operations to be performed, and the probability of each operation being performed.

Cleanroom employs *statistical usage testing* rather than coverage testing and this involves executing tests chosen from the population of all possible uses of the software in accordance with the probability of its expected use. The software reliability measure is calculated by statistical techniques based on the expected usage of the software, and Cleanroom provides a certified mean time to failure of the software.

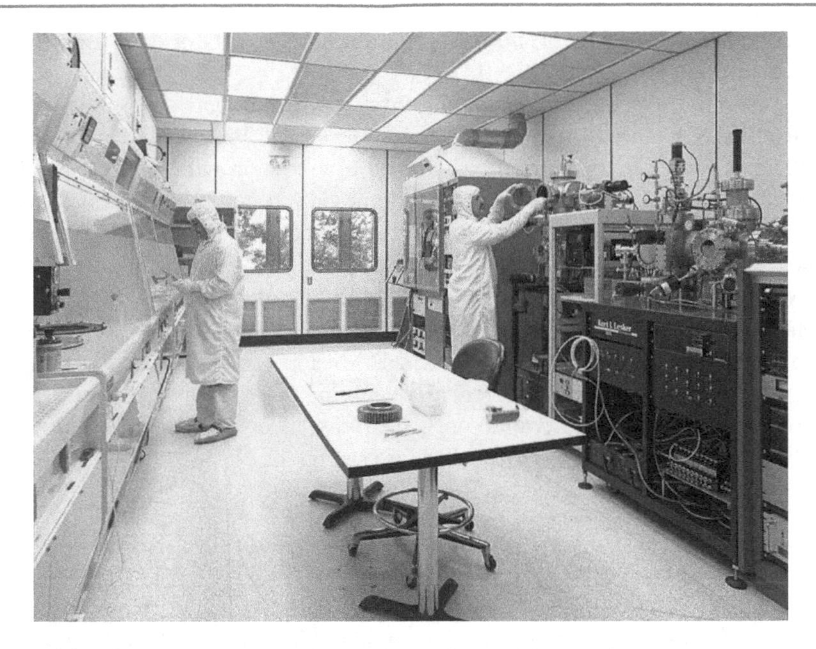

**Fig. 13.3** Cleanroom in semiconductor manufacturing. Public Domain

*Coverage testing* involves designing tests that cover every path through the program, and this type of testing is as likely to find a rare execution failure as well as a frequent execution failure. It is essential to find failures that occur on frequently used parts of the system. The advantage of usage testing (that matches the actual execution profile of the software) is that it has a better chance of finding execution failures on frequently used parts of the system. This helps to maximize the expected mean time to failure of the software.

The Cleanroom software development process is described in O'Regan (2006), and some of its successes and benefits are described in Cobb and Mills (1990). The process and calculation of the software reliability measure are described, and the Cleanroom development process enables engineers to deliver high-quality software on time and on budget.

## 13.6 Formal Methods and Testing

Formal methods have traditionally been used for the specification and development of software, but their use does not eliminate the need for software testing. Formal methods and testing are generally seen as two complementary techniques for the reduction of defects in software systems, and the development of safety-critical systems employs both techniques. A formal specification may also support testing in determining the test cases, and so formal methods may be used to improve the software testing process.

It is essential that the formal specification is correct, and so a review of the specification is required to ensure its correctness. The verification of the formal specification may take the form of specification animation with a tool, or with the use of theorem provers (usually using mechanized tools) to show the presence or absence of desirable or undesirable properties. That is, the mathematical proof is employed to show that certain desired properties are always true in the specification, whereas certain other undesirable properties are always false.

Once there is confidence in the correctness of the specification the implementation takes place, (either formal or informal development) and the system is then ready for verification with comprehensive testing. One approach where formal methods can assist is the derivation of the test cases from the formal specification, and this is termed "testing from specification" (Fig. 13.4).

## 13.7 UML and Testing

UML is an expressive graphical modelling language for visualizing, specifying, constructing, and documenting a software system. It provides several views of the software's architecture, and it has a clearly defined syntax and semantics. Each stakeholder (e.g. project manager, developers, and testers) has a different perspective and looks at the system in different ways at different times during the project. UML is a way to model the software system before implementing it in a programming language. It may be employed to document the software system, and it has been used in several domains such as the banking sector, defence, and telecommunications.

A UML specification consists of precise, complete, and unambiguous models. The models may be employed to generate code in a programming language such as Java or C++. The reverse is also possible, and so it is possible to work with either the graphical notation of UML, or the textual notation of a programming language. UML expresses things that are best expressed graphically, whereas a programming

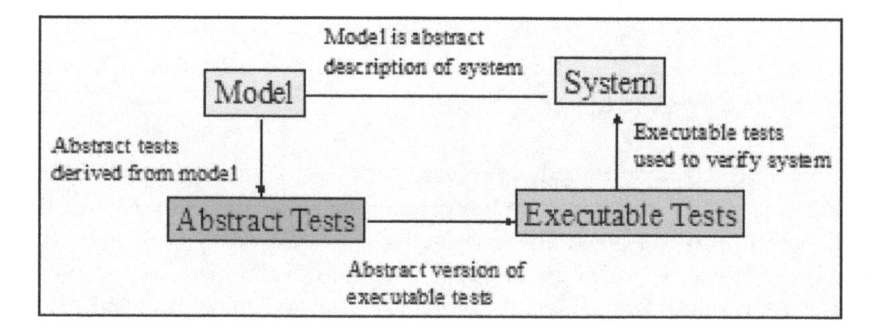


**Fig. 13.4** Deriving tests from abstract model

language expresses things that are best expressed textually, and tools are employed to keep both views consistent.

A UML model presents an abstract representation of the desired behaviour of a system under test. The test cases derived from the abstract model (the abstract test suite) is at the same level of abstraction as the model, and may not be directly executed against the system under test (Fig. 13.4). This means that the executable test suite must be derived from the abstract test suite by mapping the abstract test cases to concrete test cases that are suitable for execution.

### 13.7.1 Model Checking and Testing

Model checking is an automated technique such that given a finite-state model of a system and a formal property, (expressed in temporal logic) and then it systematically checks whether the property is true or false in a given state in the model (Fig. 13.5). It is an effective technique to identify potential design errors, and it increases the confidence in the correctness of the system design. Model checking is an effective verification technology and is widely used in the hardware and software fields. It has been employed in the verification of microprocessors; in security protocols; in the transportation sector (trains); and in the verification of software in the space sector.

Model checking is a formal verification technique based on graph algorithms and formal logic. It allows the desired behaviour (specification) of a system to be verified, and its approach is to employ a suitable model of the system and to carry out a systematic and exhaustive inspection of all states of the model to verify that the desired properties are satisfied. These properties are generally safety properties such as the absence of deadlock, request-response properties, and invariants. The

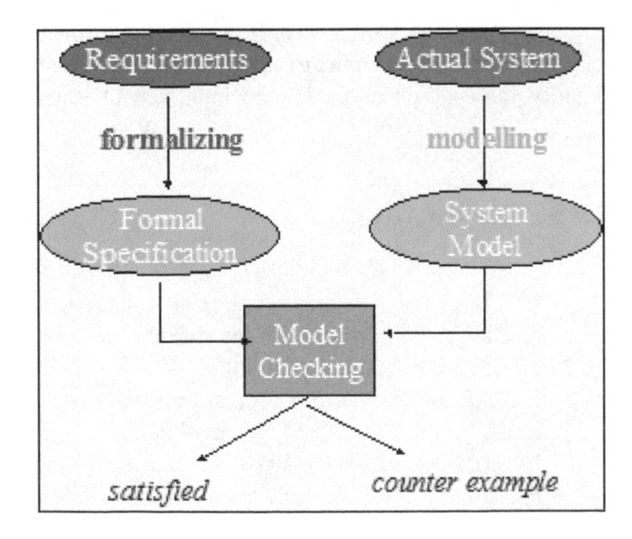


**Fig. 13.5** Model checking

**Table 13.3** Model-checking process

| Phase | Description |
|---|---|
| Modelling phase | Model the system under consideration<br>Formalize the property to be checked |
| Running phase | Run the model checker to determine the validity of the property in the model |
| Analysis phase | Is the property satisfied? If applicable, check next property<br>If the property is violated, then<br>1. Analyse generated counter example<br>2. Refine model, design or property<br>If out of space try alternative approach (e.g. abstraction of system model) |

systematic search shows whether a given system model truly satisfies a particular property or not.

The phases in the model-checking process include the modelling, running, and analysis phases (Table 13.3).

The model-based techniques use mathematical models to describe the required system behaviour in precise mathematical language, and the system models have associated algorithms that allow all states of the model to be systematically explored. Model checking is used for formally verifying finite-state concurrent systems (typically modelled by automata), where the specification of the system is expressed in temporal logic, and efficient algorithms are used to traverse the model defined by the system (in its entirety) to check if the specification holds or not. *Of course, any verification using model-based techniques is only as good as the underlying model of the system.*

Model checking is an automated technique such that given a finite-state model of a system and a formal property, and then a systematic search may be conducted to determine if the property holds for a given state in the model. The set of all possible states is called the model's state-space, and when a system has a finite state-space it is then feasible to apply model-checking algorithms to automate the demonstration of properties, with a counter example exhibited if the property is not valid. For more detailed information on model checking, see O'Regan (2019).

## 13.8 Review Questions

1. Explain the difference between software reliability and system availability
2. What is software dependability?
3. Explain the relevance of formal methods in testing
4. Describe the Cleanroom methodology
5. Describe the characteristics of a good software reliability model
6. Explain the relevance of security engineering

7. What is a safety-critical system?
8. Explain how model checking can determine whether a desired property holds at all times in a system
9. Explain how UML may support testing.

## 13.9 Summary

A safety-critical system is a system whose failure could result in significant economic damage or loss of life, and it is essential to employ rigorous processes in their design and development. Software testing alone is usually insufficient in verifying the correctness of such systems, and often an extra level of assurance is required to provide additional confidence in their correctness.

We discussed software reliability and dependability; availability; security; and safety-critical systems in this chapter. Software reliability is the probability that the program works without failure for a period of time, and it is usually expressed as the mean time to failure. Software dependability means that the software is available when required, as well as operating safely and reliably without any adverse side effects. These systems are generally fault tolerant and are designed to deal with (and recover) from faults that occur during execution.

The security of the system refers to its ability to protect itself from accidental or deliberate external attacks. There are various security threats in any networked system including threats to the confidentiality and integrity of the system and its data, and threats to the availability of the system.

Cleanroom involves the application of statistical techniques to calculate software reliability, and it is based on the expected usage of the software. Formal methods and testing are two complementary techniques, and a formal specification may also support testing in determining the test cases by deriving them from the formal specification.

A UML model presents an abstract representation of the desired behaviour of a system under test. Test cases may be derived from the abstract model (the abstract test suite), and they are at the same level of abstraction as the model. This means that the executable test suite must be derived from the abstract test suite by mapping the abstract test cases to concrete test cases suitable for execution.

Model checking is an automated technique such that given a finite-state model of a system and a formal property, (expressed in temporal logic) and then it systematically checks whether the property is true of false in a given state in the model.

## References

Bjorner D, Jones C (1982) Formal specification and software development. Prentice Hall International Series in Computer Science
Cobb RH, Mills HD (1990) Engineering software under statistical quality control. IEEE Software
Diller A (1990) An introduction to formal methods. Wiley, England
Hinchey M, Bowen J (1995) Applications of formal methods. Prentice Hall International Series in Computer Science
O'Regan G (2006) Mathematical approaches to software quality. Springer, London
O'Regan G (2019) Concise guide to formal methods. Springer, London
Spivey JM (1992) The Z Notation. A reference manual. Prentice Hall International Series in Computer Science

# 14 Legal, Ethical, and Professional Aspects of Testing

**Key Topics**

Ethics
Law of Tort
Lawsuits
Professional Responsibility
Professional Negligence
Test Outsourcing
Software Licenses
Computer Crime
Hacking

## 14.1 Introduction

Ethics is a practical branch of philosophy that deals with moral questions such as what is right or wrong, and how a person should behave in a given situation in a complex world. Ethics explore what actions are right or wrong within a specific context or within a certain society and seek to find satisfactory answers to moral questions. The origin of the word "ethics" is from the Greek word ἠθικός, which means habit or custom.

There are various schools of ethics such as the *relativist* position (as defined by Protagoras), which argues that each person decides on what is right or wrong for them; *cultural relativism* argues that the particular society determines what is right or wrong based upon its cultural values; *deontological ethics* (as defined by Kant)



G. O'Regan, *Concise Guide to Software Testing*,
Undergraduate Topics in Computer Science,
https://doi.org/10.1007/978-3-030-28494-7_14

argues that there are moral laws to guide people in deciding what is right or wrong; and *utilitarianism* (as defined by Bentham) which argues that an action is right if its overall affect is to produce more happiness than unhappiness in society.

*Professional ethics* are a code of conduct that governs how members of a profession deal with each other and with third parties. A professional code of ethics expresses ideals of human behaviour, and it defines the fundamental principles of the organization and is an indication of its professionalism. Several organizations such as the Association Computing Machinery (ACM) and the British Computer Society (BCS) have developed a code of conduct for their members, and violations of the code by members are taken seriously and are subject to investigations and disciplinary procedures.

Business ethics define the core values of the business and are used to guide employee behaviour. Should an employee accept gifts from a supplier to a company as this could lead to a conflict of interest? A company may face ethical questions on the use of technology. For example, should the use of a new technology be restricted because people can use it for illegal or harmful actions as well as beneficial ones?

Consider mobile phone technology, which has transformed communication between people, and thus is highly beneficial to society. What about mobile phones with cameras? On the one hand, they provide useful functionality in combining a phone and a camera. On the other hand, they may be employed to take indiscreet photos without permission of others, which may then be placed on inappropriate sites. In other words, how can citizens be protected from inappropriate use of such technology?

## 14.2 Business Ethics

Business ethics (also called corporate ethics) are concerned with ethical principles and moral problems that arise in a business environment (Fig. 14.1). They refer to the core principles and values of the organization and apply throughout the organization. They guide individual employees in carrying out their roles and ethical issues include the rights and duties between a company and its employees, customers, and suppliers.

Many corporation and professional organizations have a written "*code of ethics*" that defines the professional standards expected of all employees in the company. All employees are expected to adhere to these values whenever they represent the company. The human resource function in a company plays an important role in promoting ethics and in putting internal HR policies in place relating to the ethical conduct of the employees, as well as addressing discrimination, sexual harassment, and ensuring that employees are treated appropriately (including cultural sensitivities in a multi-cultural business environment).

Companies are expected to behave ethically and not to exploit its workers. There was a case of employee exploitation at the Foxconn plant (an Apple supplier of the *i*Phone) in Shenzhen in China in 2006, where conditions at the plant were so

**Fig. 14.1** Corrupt legislation. 1896. Public domain

dreadful (long hours, low pay, unreasonable workload, and crammed accommodation) that several employees committed suicide. The scandal raised questions on the extent to which a large corporation such as Apple should protect the safety and health of the factory workers of its suppliers. Further, given the profits that Apple makes from the *i*Phone, is it ethical for Apple to allow such workers to be exploited?

Today, the area of *corporate social responsibility* (CSR) has become applicable to the corporate world, and it requires the corporation to be an ethical and responsible citizen in the communities in which it operates (even at a cost to its profits). It is therefore reasonable to expect a responsible corporation to pay its fair share of tax and to refrain from using tax loopholes to avoid paying billions in taxes on international sales. Today, environment ethics has become topical, and it is concerned with the responsibility of business in protecting the environment in which it operates. It is reasonable to expect a responsible corporation to make the protection of the environment and sustainability part of its business practices.

Unethical business practices refer to those business actions that do not meet the standard of acceptable business operations, and they give the company a bad reputation. It may be that the entire business culture is corrupt or it may be result of the unethical actions of an employee. It is important that such practices be exposed, and this may place an employee in an ethical dilemma (i.e. the loyalty of the employee to the employer versus what is the right thing to do such as exposing an unethical practice).

Some accepted practices in the workplace might cause ethical concerns. For example, in many companies, it is normal for the employer to monitor email and Internet use to ensure that employees do not abuse it, and so there may be grounds for privacy concerns. On the one hand, the employer is paying the employee's salary and has a reasonable expectation that the employee does not abuse email and the Internet. On the other hand, the employee has reasonable rights of privacy provided computer resources are not abused.

The nature of privacy is relevant in the business models of several technology companies. For example, Google specializes in Internet-based services and products, and its many products include *Google Search* (the world's largest search engine); *Gmail* for email; and *Google Maps* (a Web mapping application that offers satellite images and street views). Google's products gather a lot of personal data and create revealing profiles of the users, which can then be used for commercial purposes.

A Google search leaves traces on both the computer and in records kept by Google, which has raised privacy concerns as such information may be obtained by a forensic examination of the computer, or in records obtained from Google or the Internet service providers (ISP). Gmail automatically scans the contents of emails to add context sensitive advertisements to them and to filter spam, which raises privacy concerns, as it means that all emails sent or received are scanned and read by some computer. Google has argued that the automated scanning of emails is done to enhance the user experience, as it provides customized search results, tailored advertisements, and the prevention of spam and viruses. Google's maps provide location information which may be used for targeted advertisements.

### 14.2.1 What Is Computer Ethics?

Computer ethics are a set of principles that guide the behaviour of individuals when using computer resources. Several ethical issues that may arise include intellectual property rights, privacy concerns, as well as the impacts of computer technology on wider society.

The Computer Ethics Institute (CEI) is an American organization that examines ethical issues that arise in the information technology field. It published the *ten commandments on computer ethics* (Table 14.1) in the early 1990s (Barquin 1992), which attempted to outline principles and standards of behaviour to guide people in the ethical use of computers.

The first commandment says that it is unethical to use a computer to harm another user (e.g. destroy their files or steal their personal data), or to write a program that on execution does so. That is, activities such as spamming, phishing, and cyberbullying are unethical. The second commandment is related and may be interpreted that malicious software and viruses that disrupt the functioning of computer systems are unethical. The third commandment says that it is unethical (with some exceptions such as dealing with cybercrime and international terrorism) to read another person's emails, files, and personal data, as this is an invasion of their privacy.

**Table 14.1** Ten commandments on computer ethics

| No. | Description |
|---|---|
| 1 | Thou shalt not use a computer to harm other people |
| 2 | Thou shalt not interfere with other people's computer work |
| 3 | Thou shalt not snoop around in other people's computer files |
| 4 | Thou shalt not use a computer to steal |
| 5 | Thou shalt not use a computer to bear false witness |
| 6 | Thou shalt not copy or use proprietary software for which you have not paid |
| 7 | Thou shalt not use other people's computer resources without authorization or proper compensation |
| 8 | Thou shalt not appropriate other people's intellectual output |
| 9 | Thou shalt think about the social consequences of the program you are writing or the system you are designing |
| 10 | Thou shalt always use a computer in ways that ensure consideration and respect for your fellow humans |

The fourth commandment argues that the theft or leaking of confidential electronic personal information is unethical (computer technology has made it easier to steal personal information). The fifth commandment states that it is unethical to spread false or incorrect information (e.g. fake news or misinformation spread via email or social media). The sixth commandment states that it is unethical to obtain illegal copies of copyrighted software, as software is considered an artistic or literary work that is subject to copyright. All copies should be obtained legally.

The seventh commandment states that it is unethical to break into a computer system with another user's id and password (without their permission), or to gain unauthorized access to the data on another computer by hacking into the computer system. The eight commandment states that it is unethical to claim ownership of an intellectual creation that does not belong to you. (e.g. to claim ownership of a program that was written by another).

The ninth commandment states that it is important for companies and individuals to think about the social impacts of the software that is being created and to create software only if it is beneficial to society (i.e. it is unethical to create malicious software). The tenth commandment states that communication over computers and the Internet should be courteous, as well as showing respect for others (e.g. no abusive language or spreading false statements).

### 14.2.2 The Ethical Software Tester

Software testers are professionals and need to behave ethically at all times during testing. The ISTQB code of ethics for test professionals is based on the IEEE and ACM code of ethics and it states that:

- Certified software testers shall act consistently in the public interest
- They shall act in the best interests of their client and employer
- Certified software testers shall ensure that their deliverables meet the highest professional standards
- They shall maintain independence and integrity in professional judgments
- Certified software test managers and leaders shall promote an ethical approach to the management of software testing
- They shall advance the integrity and reputation of the profession
- Certified software testers shall be supportive of their colleagues and promote cooperation with software developers
- They shall participate in lifelong learning regarding the practice of their profession and promote an ethical approach to the practice of their profession.

## 14.3 Professional Responsibility of Software Engineers and Testers

Software engineering involves multi-person construction of multi-version programs. It requires the engineer to state precisely the requirements that the software product is to satisfy and to produce designs that will meet these requirements. It involves starting with a precise description of the problem to be solved; producing a design and validating the correctness of the design; finally, the implementation and testing are performed.

Parnas has argued that computer scientists need the right education to apply scientific and mathematical principles in their work. Software engineers need education on specification, design, turning designs into programs, software inspections and testing. This should enable the software engineer to produce well-structured programs using module decomposition and information hiding. He argues that "*software engineers have individual responsibilities as professionals*"[1]. They are responsible for designing and implementing high quality and reliable software that is safe to use. They are also accountable for their own decisions and actions[2] and have a responsibility to object to decisions that violate professional standards.

Professional engineers have a duty to their clients to ensure that they are solving the real problem of the client. They need to precisely state the problem before working on its solution. Engineers need to be honest about current capabilities

[1]The concept of accountability for actions dates back thousands of years. The ancient Babylonians employed a code of laws c. 1750 B.C. known as "The Hammarabi Code". This included a law that if a house collapsed and killed the owner then the builder of the house would be executed.

[2]However, it is unlikely that an individual programmer would be subject to litigation in the case of a flaw in a program causing damage or loss of life. Most software products are accompanied by a comprehensive disclaimer of responsibility for problems (rather than a guarantee of quality).

when asked to work on problems that have no appropriate technical solution, rather than accepting a contract for something that cannot be done.[3]

The *licensing of a professional engineer* provides confidence that the engineer has the right education, experience to build safe and reliable products. Otherwise, the profession gets a bad name because of poor work carried out by unqualified people. Professional engineers are required to follow rules of good practice and to object when rules are violated. The licensing of an engineer requires that the engineer completes an accepted engineering course and understands the professional responsibility of an engineer. The professional body is responsible for enforcing standards and certification. The term "*engineer*" is a title that is awarded on merit, but *it also places responsibilities on its holder*.

Engineers have a professional responsibility and are required to behave ethically with their clients. The membership of the professional engineering body requires the member to adhere to the code of ethics of the profession. The code of ethics[4] will detail the ethical behaviour and responsibilities including (Table 14.2).

### 14.3.1 ACM Code of Professional Conduct and Ethics

The Association of Computing Machinery (ACM) has defined a code of ethics and professional conduct for its members. The general obligations are detailed in Table 14.3.

## 14.4 Legal Aspects of Testing

Legal aspects of testing are concerned with the application of the legal system to the computing field. It includes intellectual property law including patents, copyright, trademarks, and trade secrets. Patents provide legal protection for intellectual ideas; copyright law protects the expression of an idea, and trademarks provide legal protection of names or symbols. There are potential legal impacts to an organization if the software has been inadequately tested, and if the quality of the testing is deemed to be negligent.

The problem of hacking is where a hacker uses his (or her) computer skills to gain unauthorized access to a computer system. We distinguish between ethical white hat hackers and malicious black hat hackers. Computer crime includes the unauthorized access of computer resources, the theft of personal information, cyber extortion, and denial of service attacks.

[3]Parnas applied this professional responsibility faithfully when he argued against the Strategic Defence Initiative (SDI), as he believed that the public (i.e. taxpayers) were being misled and that the goals of the project were not achievable.

[4]These are core values of most mature software companies and many companies today have a code of ethics that employees are required to adhere to.

**Table 14.2** Professional responsibilities of software engineers and testers

| No. | Responsibility |
|---|---|
| 1. | Honesty and fairness in dealings with Clients |
| 2. | Responsibility for actions |
| 3. | Continuous learning to ensure appropriate knowledge to serve the client effectively |

**Table 14.3** ACM code of conduct (general obligations)

| No. | Area | Description |
|---|---|---|
| 1. | Contribute to society and human well-being | Computer professionals must strive to develop computer systems that will be used in socially responsible ways and have minimal negative consequences |
| 2. | Avoid harm to others | Computer professionals must follow best practice to ensure that they develop high-quality systems that are safe for the public. The professional has a responsibility to report any signs of danger in the workplace that could result in serious damage or injury |
| 3. | Be honest and trustworthy | The computer professional will give an honest account of their qualifications and any conflicts of interest. The professional will make accurate statement on the system and the system design and will exercise care in representing ACM |
| 4. | Be fair and act not to discriminate | Computer professionals are required to ensure that there is no discrimination in the use of computer resources, and that equality, tolerance and respect for others are respected |
| 5. | Respect property rights | The professional must not violate copyright or patent law, and only authorized copies of software should be made |
| 6. | Respect intellectual property | Computer professionals are required to protect the integrity of intellectual property, and must not take credit for another person's ideas or work |
| 7. | Respect the privacy of others | The professional must ensure that any personal information gathered for a specific purpose is not used for another purpose without the consent of the individuals. User data observed during normal system operation must be treated with the strictest confidentiality |
| 8. | Respect confidentiality | The professional will respect all confidentiality obligations to employers, clients, and users |

Software test tools are generally subject to a license, where a software license is a legal agreement between the copyright owner and the licensee that governs the use or distribution of software to the user. The two most common categories of software licenses that may be granted under copyright law are those for proprietary software and those for free open-source software.

Electronic commerce includes transactions to place an order, the acknowledgement of the order, the acceptance of the order where a legal contract now exists between both parties, and order fulfilment. We discuss the legal aspects of bespoke software development and test outsourcing, where a legal contract is prepared between the supplier and the customer. This will generally include a statement of work that stipulates the deliverables to be produced, and it may also include a service level agreement and an escrow agreement.

### 14.4.1 Legal Impacts of Failure

Software license agreements generally include limited warranties on the quality of the licensed software, and they often provide limited remedies to the customer when the software is defective. The software vendor typically promises that the software will conform to the software documentation for a specified period (the warranty period), and the software warranty generally excludes problems that are not caused by the software or are beyond the software vendor's control.

The customers are generally provided with limited remedies in the case of defective software (e.g. the replacement of the software with a corrected version, or termination of the user's right to use the defective software and a partial refund of the license fee). The payment of compensation for loss or damage is generally excluded in the software licensing agreement.

Software licensing agreements are generally accompanied by a comprehensive disclaimer that protects the software vendor from any liability (however, remote) that might result from the use of the software. It may include statements such as "*the software is provided 'as is', and that the customers use the software at their own risk*".

A limited warranty and disclaimer limit the customer's rights and remedies if the licensed software is defective, and so the customer may need to consider how best to manage the associated risks. However, there are various lawsuits that could potentially be launched against a software provider and these are discussed in the next section.

### 14.4.2 Lawsuits and Professional Negligence

A lawsuit is a proceeding by a party (or several parties) against another party in a civil court. The basic principles of litigation are where the plaintiff sues another person(s) for being negligent, and the negligence of the defendant caused injury or damage to the property of the plaintiff. It involves proving in a court of law that:

- The defendant had a duty of care
- The defendant breached this duty of care
- The breach caused harm to the plaintiff or the property of the plaintiff.

The plaintiff is entitled to compensation of the full value of the injury or the damage to the property if the case is successfully proved. Further, if there is clear evidence that the defendant acted maliciously or fraudulently then punitive damages may be awarded to the plaintiff to punish the defendant. Punitive damages are generally awarded in a small percentage of lawsuits, and they may be appealed to a higher court.

There are several types of lawsuit that may be brought against a software company (the defendant) including (Table 14.4).

### 14.4.3 The Law of Tort and Testing

The *law of tort* refers to a civil wrong where one party (the *defendant*) is held accountable for their actions (by the *plaintiff*). There are several actions that the defendant could be held accountable, e.g. negligence, trespass, misstatement, product liability, defamation, and so on. For example, the defendant may be accused of negligence and a breach of his duty of care, where damage that was reasonably foreseeable was caused by negligence.

The impact of a flaw in software may be catastrophic, and so a software development organization must take all reasonable precautions to prevent the occurrence of defects (as otherwise it may be sued for negligence). This is especially true in the safety-critical domain, where defects could cause major damage or even loss of life. Reasonable precautions consist of having appropriate software

**Table 14.4** Types of lawsuits

| Type | Description |
|---|---|
| Criminal | This type of lawsuit is brought by the state against the software company (or developers or testers) for committing a criminal act (e.g. tampering with a computer or loading a virus onto a computer) |
| Tort | This type of lawsuit is brought by an individual(s) against a company/developers for committing some wrong to you or your computer (e.g. releasing a virus onto your computer) |
| Negligence | The company has a duty of care to take reasonable measures to make the product safe, so that there are no personal injuries or damage to property |
| Malpractice | This is where the quality of service is judged against a professional standard and deemed to be negligent, with mistakes made in the delivery of the service that would not be made by an ordinary professional in the field |
| Strict liability | A product defect caused a personal injury or damage to property, and the burden of proof required is to demonstrate that the program was defective and that the defect caused the accident (e.g. failure of program controlling breaks in a car) |
| Fraud | The company made a statement of fact to you when it knew that the statement was false (and where you relied on the statement to make an economic decision such as buying a defective product) |
| Regulatory | The regulatory sector (e.g. FDA) places requirements on how software should be developed and tested so that it is safe for the public to use |
| Breach of contract | A software contract specifies the obligations that both parties have to each other (as well as implied terms such as implied warranty) |

engineering practices in place to allow the organization to consistently produce high-quality software.

A quality management system indicates that the organization takes software quality seriously and has a sound software development process in place that serves the needs of the organization and its customers. Modem quality assurance systems include processes for software inspections, testing, quality audits, customer satisfaction, software development, project planning, etc.

The organization will require evidence or records to prove that the quality management system is in place, and that it is appropriate for the organization, and that it is fully operational within the organization. This generally requires records and an audit trail of the various quality activities to be maintained. The records enable the organization to prepare a legal defence to show that it took all reasonable precautions in software development, especially if a customer decides to take legal action for negligence against the software provider following a serious problem in the software at the customer site.

The presence of records may be used to indicate that all reasonable steps were taken, and the records typically include lists of all the deliverables in the project; minutes of project meetings; records of reviews of requirements, design, and software code, records of test plans and test results; and so on.

## 14.5 Legal Aspects of Test Outsourcing

Test outsourcing and bespoke software development have become popular in the software engineering field. Test outsourcing is where the testing is outsourced to an independent external organization. Bespoke (or custom) software is software that is developed for a specific customer or organization, and it needs to satisfy the defined customer requirements. The organization will need to be rigorous in its selection of the appropriate supplier (as discussed in Chap. 8), as it is essential that the supplier selected has the capability of delivering high quality and reliable software on time and on budget.

This means that the capability of the supplier is clearly understood and the associated risks are known prior to selection. The selection is based on objective criteria such as cost, the approach, the ability of the supplier to deliver the required solution, the supplier capability, and while cost is an important criterion, and it is just one among several other important factors.

Once the selection of the supplier is finalized a legal agreement is drawn up between the contractor and supplier, which states the terms and condition of the contract, as well as the statement of work (Fig. 14.2). The *statement of work* (SOW) details the work to be carried out, the deliverables to be produced, when they will be produced, the personnel involved their roles and responsibilities, any training to be provided, and the standards to be followed. The agreement will need to be signed by both parties and may (depending on the type of agreement) include:

**Fig. 14.2** Legal contract. Creative Commons

- Legal contract
- Statement of work
- Implementation plan
- Training plan
- User guides and manuals
- Customer support to be provided
- Service level agreement
- Escrow agreement
- Warranty period.

A *service level agreement* (SLA) is an agreement between the customer and service provider, which specifies the service that the customer will receive as well as the response time to customer issues and problems. It will also detail the penalties should the service performance fall below the defined levels.

An *escrow agreement* is an agreement made between two parties where an independent trusted third party acts as an intermediary between both parties. The intermediary receives money from one party and sends it to the other party when contractual obligations are satisfied. Under an escrow agreement the trusted third party may also hold documents and source code.

Occasionally, it will be just the testing part of a project that is outsourced, and test outsourcing is concerned with the selection and management of an appropriate supplier to perform the testing. It is essential that the selected test organization is capable of carrying out the required testing to the defined quality standard, as well as being capable of completing the testing within the budget and schedule constraints.

The legal contract specifies the obligations of the supplier and should the supplier fail to honour its commitments it may well be in breach of contract. This means that the binding agreement has not been honoured, and there may be a need to seek legal remedy if a *material* breach of the contract has occurred. The first step is dialogue between both parties with the objective of finding a reasonable resolution, but if both parties are unable to agree a way forward the first party may seek a legal remedy in a civil court.

## 14.6 Licenses for Test Tools

Testers often employ dedicated test tools for various parts of the test process, and the use of tools is generally subject to a licensing agreement. The tools may be developed in-house, but it is more common to employ proprietary tools or open-source tools. A software license is a legal agreement between the copyright owner and the licensee, which governs the use or distribution of software to the user (licensee). Computer software code is protected under copyright law in most countries, and a typical software license grants the user permission to make one or more copies of the software, where the copyright owner retains exclusive rights to the software under copyright law.

The two most common categories of software licenses that may be granted under copyright law are those for *proprietary software* and those for *free open-source software* (FOSS). The rights granted to the licensee are quite different for each of these categories, where the user has the right to copy, modify, and distribute (under the same license) software that has been supplied under an open-source license, whereas proprietary software typically does not grant these rights to the user.

The *licensing of proprietary software* typically gives the owner of a copy of the software the right to use it (including the rights to make copies for archival purposes). The software may be accompanied with an end-user license agreement (EULA) that may place further restrictions on the rights of the user. There may be restrictions on the ownership of the copies made, and on the number of installations allowed under the term of the distribution. The ownership of the copy of the software often remains with the copyright owner, and the end user must accept the license agreement to use the software.

The most common licensing model is per single user, and the customer may purchase a certain number of licenses over a fixed period. Another model employed is the license per server model (for a site license), or a license per dongle model, which allows the owner of the dongle use the software on any computer. A license may be perpetual (it lasts forever), or it may be for a fixed period (typically one year).

The software license often includes maintenance for a period (typically one year), and the maintenance agreement generally includes updates to the software during that time and it may also cover a limited amount of technical support. The two parties may sign a service level agreement (SLA), which stipulates the service that will be provided by the service provider. This will generally include timelines for the resolution of serious problems, as well as financial penalties that will be applicable where the customer service performance does not meet the levels defined in the SLA.

Free- and open-source licenses are often divided into two categories depending on the rights to be granted in distribution of the modified software. The first category aims to give users unlimited freedom to use, study, and modify the software, and if the user adheres to the terms of an open-source license such as the Free Software Foundation (FSF), GNU or General Public License (GPL), the freedom to distribute the software and any changes made to it. The second category of open-source licenses give the user permission to use, study, and modify the software, but not the right to distribute it freely under an open-source license (it could be distributed as part of a proprietary software license).

## 14.7 Testing and Prevention of Computer Crime

It is common in the major urban areas to encounter dangers in some streets or neighbourhoods, and such dangers need to be managed. Similarly, the Internet has dangers with hackers, scammers, and Web predators lurking in the shadows. A hacker may be accessing a computer resource without authorization with the intention of committing an unlawful act. The hacker's activities may be limited to *eavesdropping* (listening to a conversation), or it may be an active *man-in-the-middle* attack, where the hacker may possibly alter the conversation between two parties.

One of the earliest Internet attacks was back in 1988 when a graduate student from Carnegie Mellon University released a program on the Internet (an Internet Worm) that exploited security vulnerability in the mail software to automatically replicate itself locally and on remote machines. It affected lots of machines and effectively shut down the Internet for 1–2 days.

Today, more and more individuals and companies are on line, and networking systems and computers have become quite complex. There has been a major growth in attacks on businesses and individuals, and so it is essential to consider computer and network security. The Internet was developed based on trust with security features added as a response to different types of attacks.

There are several threats associated with network connectivity such as *unauthorized access* (a break-in by an unauthorized person), *disclosure of sensitive information* to people who should not have access to the information, and *denial of service* (DoS), where there is a degradation of service that makes it impossible to access the Web site and perform productive work.

There may be attacks that lead to defacement of the Web sites, bank fraud, theft of credit card numbers, hoax (scam) letters, and phishing emails that appear to come from legitimate parties but contain links to a site that is different from the one that the user expects to go to, intercepting of packets and password sniffing. *Phishing* is an attempt to obtain sensitive information such as usernames, passwords, and credit card details with the intention of committing fraud.

A computer *virus* is a self-replicating computer program that is installed on the user's computer without consent. It is a malicious software program that when executed replicates itself and infects other computer programs by modifying them. A virus often performs some type of harmful activity on the infected computers such as accessing private information, spamming email contacts, or corrupting data. It is not a crime per se to write a computer virus or malicious software. However, if that software or other malware spreads to other computers, then it could be considered a crime.

*Cyberextortion* is a crime that involves an attack, or threat of an attack, accompanied by a demand for money to stop the attack. They are often initiated through malware in an email attachment. These may include denial of service attacks or *ransomware* attacks that encrypts the victim's data. The victim is then offered the private key to resolve the encryption in return for payment. Companies need to manage the risks associated with cyberextortion and to ensure that end users are properly educated on malware and phishing.

Another form of computer crime is Internet fraud where one party is intent on deceiving another. Among these are hoax email scams, which are designed to deceive and fraud the email recipient. These may include the *Nigeria 419* scams, where the email recipient is offered a share of a large amount of money trapped in their country, if the recipient will help in getting the money out of the country. The recipient may be asked for their bank account details to help them to transfer the money (this information will later be used by them to steal funds), or the request may be to pay fees or taxes to release payment with further fees requested. Of course, the money will never arrive (*if an email looks like it really is too good to be true then it has a high probability of being a scam*).

Security testing of the software is important, as it is essential to identify any security vulnerabilities and to correct them. Further, it is important that users be educated to minimize risks of becoming victims of computer crime.

### 14.7.1 Testing and Hacking

A *hacker* is a person who uses his (or her) computer skills to gain unauthorized access to computer files or networks. A hacker may enjoy experimenting with computer technology (the original meaning of the term), but some hackers enjoy breaking into systems and causing damage (the modern meaning of the word). Ethical (*white hat*) hackers are former hackers who play an important role in the security industry in testing network security and in helping to create secure products and services. Malicious (*black hat*) hackers (also called *crackers*) are generally motivated by personal gain, and they exploit security and system vulnerabilities to steal, exploit or sell data (Fig. 14.3).

Many computer systems in use today have vulnerabilities that may be exploited by a determined hacker to gain unauthorized entry to the system and access to unauthorized information. It is vital that best practice in software and system engineering is employed to develop safe and secure systems, and that known vulnerabilities in system security are addressed promptly by updates to the system software. Further, it is essential to educate staff on security and to define (and follow) the appropriate procedures to prevent security breaches.

The early hackers were mainly young students without malicious intent who were exploring the university computer systems. These include the students at Massachusetts Institute of Technology in the late 1950s who were interested in exploring the IBM 704 computer, and they would enter areas of the system without authorization and gain access to privileged resources. They were motivated by knowledge and wished to have a deeper understanding of the systems that they had access to. The idea of a hacker ethic was formulated in a book by Steven Levy in the mid-1980s (Levy 1984), and he outlined several ethical principles including free access to computers and information and improvement to quality of life. His six key tenets are:

**Fig. 14.3** Hacker at work on backlit keyboard. Creative Commons

– Access to computers should be unlimited and total
– All information should be free
– Mistrust authority
– Hackers should be judged by their hacking and not by bogus criteria such as race and religion
– Art and beauty can be created on a computer
– Computers can change your life for the better.

The *free software movement* arose in the early 1980s from followers of the hacker ethic, with Richard Stallman (its founder) often referred to as "the last true hacker" (O'Regan 2015). Today, ethical hackers need to obtain permission prior to acting, as their actions may potentially cause major disruption to an organization. Responsible (white hat) hackers can provide useful information on security vulnerabilities, and may assist by testing and improving computer security.

The security of the system refers to its ability to protect itself from accidental or deliberate external attacks, which are common today since most computers are networked and connected to the Internet. There are various security threats in any networked system including threats to the confidentiality and integrity of the system and its data, and threats to the availability of the system.

Therefore, controls are required to enhance security and to ensure that attacks are unsuccessful. Encryption is one way to reduce system vulnerability, as encrypted data is unreadable to the attacker. There may be controls that detect and repel attacks, and these controls are used to monitor the system and to take appropriate action to shut down parts of the system or restrict access in the event of an attack. There may be controls that limit exposure (e.g. insurance policies and automated backup strategies) that allow recovery from the problems introduced.

The introduction of the Internet in the early 1990s has transformed the world of computing, and it later led to an explosive growth in attacks on computers and systems, as hackers and malicious software sought to exploit known security vulnerabilities. It is therefore essential to develop secure systems that can deal with and recover from such external attacks.

Hackers will often attempt to steal confidential data and to disrupt the services being offered by a system. Security engineering is concerned with the development of systems that can prevent such malicious attacks, and recover from them. It has become an important part of software and system engineering, and software developers need to be aware of the threats facing a system and develop solutions to manage them.

Hackers may probe parts of the system for weaknesses, and system vulnerabilities may lead to attackers gaining unauthorized access to the system. There is a need to conduct a risk assessment of the security threats facing a system early in the software development process, and this will lead to several security requirements for the system.

The system needs to be designed for security, as it is difficult to add security after the system has been implemented. Security loopholes may be introduced in the development of the system, and so care needs to be taken to prevent these as well as preventing hackers from exploiting security vulnerabilities.

The choice of architecture and how the system is organized is fundamental to the security of the system, and different types of systems will require different technical solutions to provide an acceptable level of security to its users. The following guidelines for designing secure systems are described in Sommerville (2011):

- Security decisions should be based on the security policy
- A security-critical system should fail securely
- A secure system should be designed for recoverability
- A balance is needed between security and usability
- A single point of failure should be avoided
- A log of user actions should be maintained
- Redundancy and diversity should be employed
- Organization of information in system into compartments.

Security testing is carried out to identify any flaws in the security mechanisms of the computer system and to verify that the security requirements such as confidentiality, availability, integrity, etc., are satisfied. However, the successful completion of security testing does not guarantee that there are no security vulnerabilities in the system.

The unauthorized access to a computer system and the theft of confidential data and disruption of its services is unlawful and may be subject to prosecution and the full rigour of the law.

## 14.8 Review Questions

1. What is intellectual property law?
2. Describe the behaviours of the ethical software tester
3. How can a software company demonstrate that it took all reasonable steps to deliver a high-quality software product, and that the testing was fit for purpose
4. Explain the different types of software licensing
5. Explain the legal aspects of bespoke software development
6. What happens when one party in a test-outsourcing project believes that a material breach of the contract has occurred?

7. What types of lawsuits could be brought against a software company?
8. Explain the difference between ethical and malicious hackers
9. What is computer crime?
10. Explain cyber extortion.

## 14.9 Summary

Legal aspects of testing are concerned with the application of the legal system to the computing field. It includes intellectual property law including patents, copyright, trademarks and trade secrets; bespoke software development; test outsourcing; licensing of software; professional negligence in the development and testing of software; and computer crime.

A lawsuit is a proceeding by a party against another party in a civil court where the plaintiff sues another person for being negligent, and the negligence of the defendant caused injury or damage to the property of the plaintiff.

Bespoke software (or custom software) is software that is developed for a specific customer or organization and needs to satisfy specific customer requirements. The legal contract specifies the obligations of the supplier, and should the supplier fail to honour its commitments it may well be in breach of contract. This may result in the first party seeking a legal remedy in a civil court.

A software license is a legal agreement between the copyright owner and the licensee, which governs the use or distribution of software to the user (licensee). Computer software code is protected under copyright law, and the license grants the user permission to make one or more copies of the software. Software license agreements generally provide limited remedies to the customer when the software defective. However, there may be legal implications if the software has been inadequately developed and tested.

A hacker is a person who uses his (or her) computer skills to gain unauthorized access to computer files or networks. Hackers may probe parts of the system for weaknesses, and system vulnerabilities may lead to attackers gaining unauthorized access to the system. The system needs to be designed for security, as it is difficult to add security after the system has been implemented. Security loopholes may be introduced in the development of the system, and so care needs to be taken to prevent these as well as preventing hackers from exploiting security vulnerabilities.

## References

Barquin RC (1992) In pursuit of a 'ten commandments' for computer ethics. Computer Ethics Institute, Washington, D.C.
Levy S (1984) Hackers: heroes of the computer revolution. O'Reilly Media, Sebastopol
O'Regan G (2015) Pillars of computing. Springer, Berlin
Sommerville I (2011) Software engineering, 9th edn. Pearson, London

# Configuration Management 15

**Key Topics**

Configuration management system
Configuration items
Baseline
File naming conventions
Version control
Change control
Change control board
Configuration management audits

## 15.1 Introduction

Software configuration management (SCM) is concerned with managing and controlling changes to the software and project deliverables, and it provides full traceability of the changes made during the project. It provides a record of what has been changed, as well as who changed it. SCM involves identifying the configuration items of the system; controlling changes to them; and maintaining integrity and traceability.

The origins of software configuration management go back to the early days of computing, when the principles of configuration management used in hardware design and development were applied to software development in the 1950s. It has evolved over time to a set of procedures and tools to manage changes to the software.



G. O'Regan, *Concise Guide to Software Testing*,
Undergraduate Topics in Computer Science,
https://doi.org/10.1007/978-3-030-28494-7_15

The configuration items are generally documents in the early part of the software development lifecycle, whereas the focus is on source code control management and software release management in the later parts of development. Software configuration management involves:

– Identifying what needs to be controlled
– Ensuring those items are accurately defined and documented
– Ensuring that changes are made in a controlled manner
– Ensuring that the correct version of a work product is being used
– Knowing the version and status of a configuration item at any time
– Ensuring adherence to standards
– Planning builds and releases.

Software configuration management allows the orderly development of software, and it ensures that only authorized changes to the software are made. It ensures that releases are planned, and that the impacts of proposed changes are considered prior to their authorization. The integrity of the system is maintained at all times, and the constituents of the software (including their version numbers) are known at any time.

Effective configuration management allows questions such as the following (Table 15.1) to be easily answered:

The symptoms of poor configuration management include corrected defects that suddenly begin to reappear; difficulty in or failure to locate the latest version of source code; or failure to determine the source code that corresponds to a software release.

Therefore, it is important to employ sound configuration management practices to enable high-quality software to be consistently produced. Poor configuration management practices lead to quality problems resulting in a loss of the credibility and reputation of a company. Several symptoms of poor configuration management practices are listed in Table 15.2.

Configuration management involves identifying the configuration items to be controlled, and systematically controlling change to them, in order to maintain the integrity and traceability of the configuration throughout the software development lifecycle. There is a need to manage and control changes to documents and source

**Table 15.1** Features of good configuration management

| Features of good configuration management |
|---|
| What is the correct version of the software module to be updated? |
| Where can I get a copy of R4.7 of Software System X? |
| What versions of the Software System X are installed at the various customer sites? |
| What changes have been introduced in the new release of software (version R4.8 from the previous release of R4.7)? |
| What version of the design document corresponds to software system version R3.5? |
| What customers use R3.5 of the software system? |
| Are there undocumented or unapproved changes included in the released version of the software? |

**Table 15.2** Symptoms of poor configuration management

| Symptoms of poor configuration management |
|---|
| Defects corrected suddenly begin to reappear |
| Cannot find the latest version of the source code |
| Unable to match the source code and object code |
| Wrong version of software sent to the customer |
| Wrong code tested |
| Cannot replicate previously released code |
| Simultaneous changes to same source component by multiple developers with some changes lost |

code, including the project plan, the requirements document, design documents, code, and test plans.

A key concept in configuration management is that of a "*baseline*", which is *a set of work products that have been formally reviewed and agreed upon, and serves as the foundation for future development work.*

A baseline can only be changed through formal change control procedures, which leads to a new baseline. It provides a stable basis for the continuing evolution of the configuration items, and all approved changes move forward from the current baseline leading to the creation of a new baseline. The change control board (CCB) or a similar mechanism authorizes the release of baselines, and the content of each baseline is documented. All configuration items must be approved before they are entered into the released baselines.

Therefore, it is necessary to identify the configuration items that need to be placed under formal change control, and to maintain a history of the changes made to the baseline. There are four key parts to software configuration management (Table 15.3).

A typical set of software releases (e.g. in the telecommunications domain) consists of incremental development, where the software to be released consists of a number of release builds with the early builds consisting of new functionality, and the later builds consisting of fix releases.

Software configuration management is planned for the project, and each project will typically have a configuration management plan, which will detail the planned delivery of functionality and fix releases for the project (Table 15.4).

Each of the R.1.0.0.*k* baselines are termed release builds, and they consist of new functionality and fixes to the identified problems. The content of each release build is known, i.e. the project team and manager will target specific functionality and fixes for each build, and the actual content of the particular release baseline is documented. Each release build can be replicated, as the version of source code to create the build is known, and the source code is under control management.

There are various tools employed for software configuration management activities, including well-known tools such as ClearCase, PVCS, and Visual Source Safe (VSS) for source code control management. The PV tracker tool and ClearQuest may be used for tracking defects and change requests. A defect-tracking tool will list all of the open defects against the software, and a defect may require several

**Table 15.3** Software configuration management activities

| Area | Description |
|---|---|
| Configuration identification | This requires identifying the configuration items to be controlled, and implementing a sound configuration management system, including a repository where documents and source code are placed under controlled access. It includes a mechanism for releasing documents or code, a file naming convention and a version numbering system for documents and code, and baseline/release planning. The version and status of each configuration item should be known |
| Configuration control | This involves tracking and controlling change requests and controlling changes to the configuration items. Any changes to the work products are controlled, and authorized by a change control board or similar mechanism. Problems or defects reported by the test groups or customer are analysed, and any changes made are subject to change control. The version of the work product is known and the constituents of a particular release are known and controlled. The previous versions of releases can be recreated, as the source code constituents are fully known and available |
| Configuration auditing | This includes audits to verify the integrity of the baseline, and audits of the configuration management system verify that the standards and procedures are followed. The results of the audits are communicated to the affected groups, and corrective action taken to address the findings |
| Status accounting | This involves data collection and report generation. These reports include the software baseline status, the summary of changes to the software baseline, problem report summaries, and change request summaries |

**Table 15.4** Build plan for project

| Release Baseline | **Contents** | **Date** |
|---|---|---|
| R 1.0.0.0 | $F_4$, $F_5$, $F_7$ | 31.01.17 |
| R. 1.0.0.1 | $F_1$, $F_2$, $F_6$+ fixes | 15.02.17 |
| R. 1.0.0.2 | $F_3$+ fixes | 28.02.17 |
| R. 1.0.0.3 | $F_8$ + fixes (functionality freeze) | 07.03.17 |
| R. 1.0.0.4 | Fixes | 14.03.17 |
| R. 1.0.0.5 | Fixes | 21.03.17 |
| R. 1.0.0.6 | Official release | 31.03.17 |

change requests to correct the software (as a problem may affect different parts of the software product, as well as different versions of the product, and a change request may be necessary for each part). The tool will generally link the change requests to the problem report. The current status of the problem report can be determined, and the targeted release build for the problem identified.

The CMMI provides guidance on practices to be implemented for sound configuration management (Table 15.5).

**Table 15.5** CMMI requirements for configuration management

| Specific goal | Specific practice | Description of specific practice/goal |
|---|---|---|
| SG 1 | | **Establish baselines** |
| | SP 1.1 | Identify configuration items |
| | SP 1.2 | Establish a configuration management system |
| | SP 1.3 | Create or release baselines |
| SG 2 | | **Track and control changes** |
| | SP 2.1 | Track change requests |
| | SP 2.2 | Control configuration items |
| SG 3 | | **Establish integrity** |
| | SP 3.1 | Establish configuration management records |
| | SP 3.2 | Perform configuration audits |

The CMMI requirements are concerned with establishing a configuration management system; identifying the work products that need to be subject to change control; controlling changes to these work products over time; controlling releases of work products; creating baselines; maintaining the integrity of baselines; providing accurate configuration data to stakeholders; recording and reporting the status of configuration items and change requests; and verifying the correctness and completeness of configuration items with configuration audits. We shall discuss the key parts of configuration management in the following sections.

## 15.2 Configuration Management System

The configuration management system enables the controlled evolution of the documents and the software modules produced during the project. It includes

- Configuration management planning
- A document repository with check in/check out features
- A source code repository with check in/check out features
- A configuration manager (may be a part time role)
- File naming convention for documents and source code
- Project directory structure
- Version numbering system for documents
- Standard templates for documents
- Facility to create a baseline
- A release procedure
- A group (change control board) to approve changes to baseline
- A change control procedure
- Configuration management audits to verify integrity of baseline.

### 15.2.1 Identify Configuration Items

The configuration items are the work products to be placed under configuration management control, and they include project documents, source code, and data files. They may also include compilers as well as any supporting tools employed in the project.

The project documentation will typically include project plans; the user requirements specification; the system requirements specification; the architecture and technical design documents; the test plans, etc.

The items to be placed under configuration management control are identified and documented early in the project lifecycle. Each configuration item needs to be uniquely identified and controlled. This may be done with a naming convention for the project deliverables and source code, and applying it consistently. For example, a simple approach is to employ mnemonics labels and version numbers to uniquely identified project deliverables. A user requirements specification for project 005 in the Finance business area may be represented simply by:

FIN_005_URS.

### 15.2.2 Document Control Management

The project documents are stored in a document repository using a configuration management tool such as PVCS or VSS. For consistency, a standard directory structure is often employed for projects, as this makes it easier to locate particular configuration items. A single repository may be employed for both documents and software code (or a separate repository for each).

Clearly, it is undesirable for two individuals to modify the same document at the same time, and the document repository will include *check in/check out* procedures. The document must be checked out prior to its modification, and once it is checked out, another user may not modify it until it has been checked back in. An audit trail of all modifications made to a particular document is maintained, including details of the person who made the change, the date that the change was made, and the rationale for the change.

**Version Numbering of Documents**

A simple version numbering system may be employed to record the versions of documents: e.g. v0.1, v0.2, and v0.3 is often used for draft documents, with version v1.0 being the first approved version of the document. Each time a document is modified its version number is incremented, and the document history records the reasons for modification.

- V0.1 Initial draft of document
- V0.$x$ Revised draft ($x > 0$)
- V1.0 Approved baseline version

– V1.$x$ Approved minor revision ($x > 0$)
– V$n$.0 Approved major revision ($n > 1$)
– V$n$.$x$ Approved minor revision ($x > 0$, $n > 1$).

The document will provide information on whether it is a draft or approved, as well as the date of last modification, the person who made the modification, and the rationale for the modification. The configuration management system will provide records of the configuration management activities, as well as the status of the configuration items and the status of the change requests. The revision history of the configuration items will be maintained.

### 15.2.3 Source Code Control Management

The source code and data files are stored in a source code repository using a tool such as PVCS, VSS or Clearcase, and the repository provides an audit trail of all the changes made to the source code. An item must first be checked out for modification, the changes are made, and it is then checked back into the repository. The source code management system provides security and control of the configuration items, and the procedures include:

– Access controls
– Checking in/out configuration items
– Merging and branching
– Labels (labelling releases)
– Reporting.

The source code configuration management tool ensures the integrity of the source code, and prevents more than one person from altering the software code at the same time.

### 15.2.4 Configuration Management Plan

A software *configuration management plan* (it may be part of the project plan or a separate plan) is prepared early in the project, and it defines the configuration management activities for the project. It will detail the items to be placed under configuration management control, the standards for naming configuration items, the version numbering system, as well as version control and release management.[1] The CM plan is placed under configuration management control.

The content of each software release is documented, as well as installation and rollback instructions. The content includes the requirements and change requests

[1]These may be defined in a configuration management procedure and referenced in the CM plan.

implemented, as well as the defects corrected and the version of the new release. A list is maintained of the customer sites of where the release has been installed. All software releases are tested prior to their approval. The CM plan will include:

- Roles and responsibilities
- Configuration items
- Naming conventions
- Version control
- Filing structure for project.

The stakeholders and roles involved are identified and documented in the CM plan. Often, the role of a software configuration manager is employed, and this may be a full time or part time role.[2] The CM manager ensures that the configuration management activities are carried out correctly, and will conduct and report the results of the CM audits.

## 15.3 Change Control

A change request (CR) database[3] is set-up to record change requests made during the project. The change requests are documented and considered by the change control board (CCB). The CCB may just consist of the project manager and the system owner for small projects, or a management and technical team for larger projects.

The impacts and risks of the proposed change need to be considered, and an informed decision made on whether to reject or approve the CR. The proposed change may have technical impacts, as well as introducing new project risks, and may adversely affect the schedule and budget. It is important to keep change to a minimum at the later stages of the project in order to reduce risks to quality.

Figure 15.1 describes a simple process for raising a change request; performing an impact assessment; deciding on whether to approve or reject the change request; and proceeding with implementation (where applicable).

The results of the CCB review of each change request (including the rationale of the decision made) will be recorded. Change requests and problem reports for all configuration items are recorded and analysed, reviewed, approved (or rejected), and tracked to closure.

A sample configuration management process map is detailed in Fig. 15.2, and it shows the process for updates to configuration information following an approved change request. The deliverable is checked out of the repository; modifications are made and the changes approved; configuration information is updated; and the deliverable is checked back into the repository.

[2]This depends on the size of the organization and projects. The project manager may perform the CM manager role for small projects.
[3]This may just be a simple Excel spread sheet or a sophisticated tool.

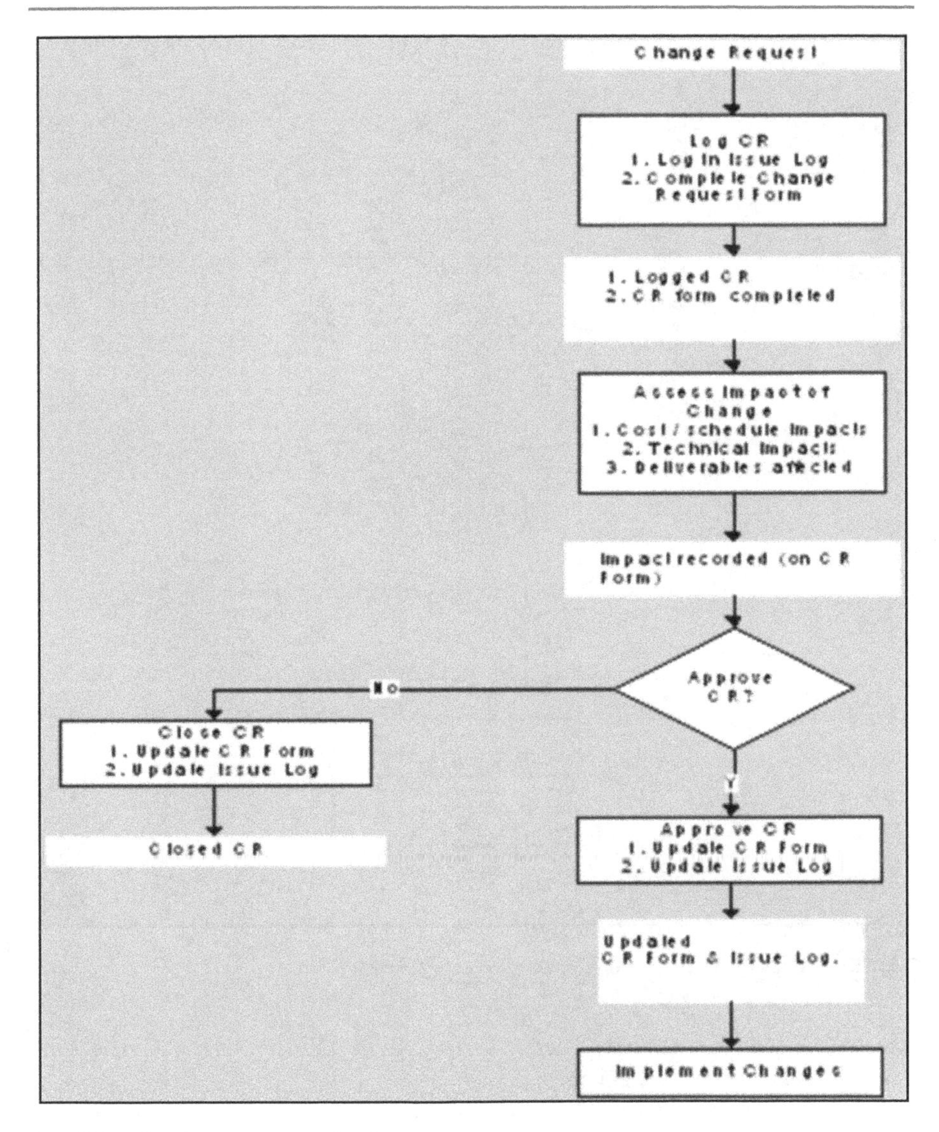


**Fig. 15.1** Simple process map for change requests

## 15.4 Configuration Management Audits

*Configuration management audits* are conducted during the project to verify that the configuration is consistent and complete. Every project should have at least one configuration audit, and the objective is to verify the completeness and correctness of the configuration system for the project. The audit will check that the records correctly identify the configuration, and that the configuration management

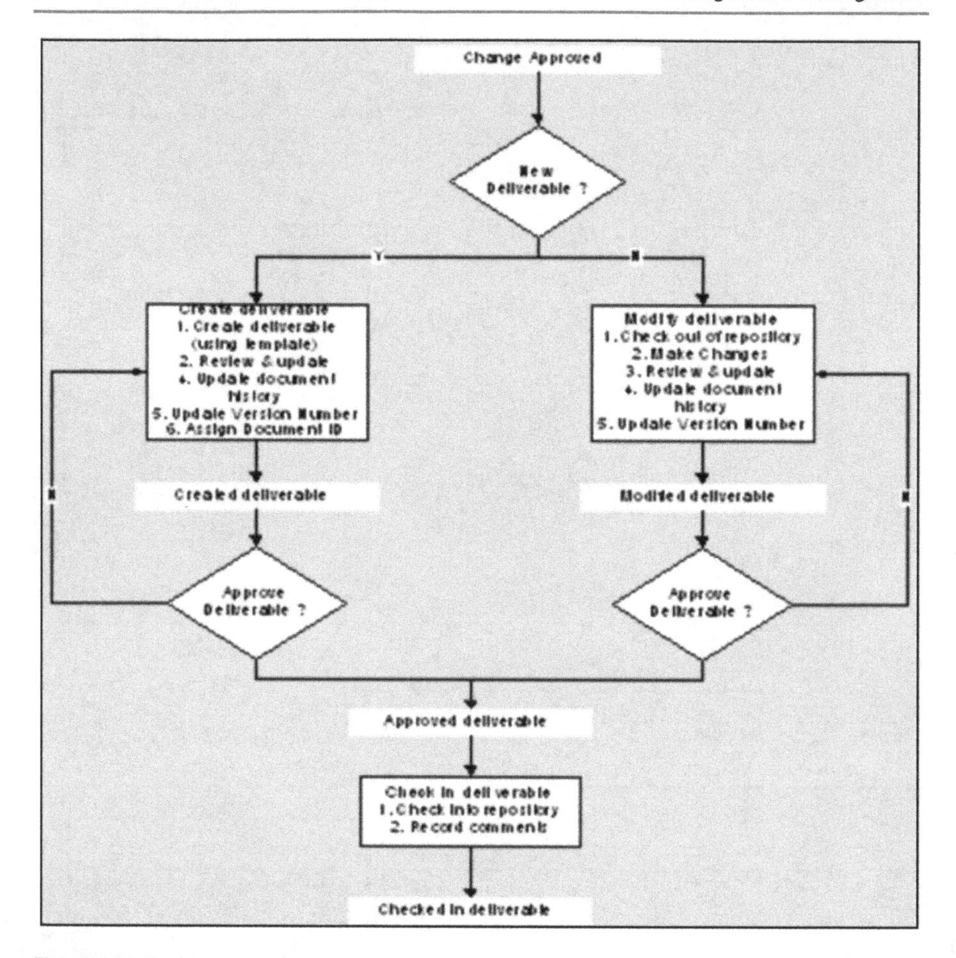


**Fig. 15.2** Simple process map for configuration management

standards and procedures have been followed. Table 15.6 presents a sample configuration management checklist.

There may also be a *librarian role* to set-up the filing structure for the project, or the configuration manager may perform this role. The project manager assigns responsibilities for performing configuration management activities. All involved in the process receive appropriate training on the process.

## 15.5 Configuration Management in Testing

Configuration management in testing is concerned with managing the configuration of the Testware. This includes placing all of the testing deliverables produced during the testing (i.e. during planning, designing, and executing tests) under

**Table 15.6** Sample configuration management audit checklist

| No. | Item to check |
|---|---|
| 1. | Is the directory structure set-up for the project? |
| 2. | Are the configuration items identified and listed? |
| 3. | Have the latest versions of the templates been used? |
| 4. | Is a unique document Id employed for each document? |
| 5. | Is the standard version numbering system followed for the project? |
| 6. | Are all versions of documents and software modules in the document/source code repository? |
| 7. | Is the configuration management plan up to date? |
| 8. | Are the roles defined in the configuration management plan performing their assigned responsibilities? |
| 9. | Are changes to the approved documents formally controlled? |
| 10. | Is the version number of a document incremented following an agreed change to an approved document? |
| 11. | Is there a change control board set-up to approve change requests? |
| 12. | Is there a record of which releases are installed at the various customer sites? |
| 13. | Are all documents/software modules produced by vendors under appropriate configuration management control? |

configuration management control. This includes documentation such as test plans, test specifications, and test reports; test scripts, test environments, test tools; databases and the test results.

The configuration management of testing ensures that there is an up-to-date record of what has been tested, including the versions of the underlying files and the components from which the software has been built. The developers prepare a version-controlled test release from a well-managed source code control repository. The testers raise defect reports against the version of software that was tested. It is then possible for the developers to identify the versions of the source files that need to be modified to correct the defect, and to make the necessary corrections. The developers then prepare a new release for the testers to verify.

## 15.6 Review Questions

1. What is software configuration management?
2. What is change control?
3. What is a baseline?
4. Explain source code control management.
5. Explain document control management.

6. What is a configuration management audit and explain how it differs from a standard audit?
7. Describe the role of the configuration manager and librarian.
8. Describe the main elements in a software configuration management system.

## 15.7 Summary

Software configuration management is concerned with the orderly development and evolution of the software. It is concerned with tracking and controlling changes to the software and project deliverables, and it provides full traceability of the changes made during the project.

It involves identifying the configuration items that are subject to change control, controlling changes to them, and maintaining integrity and traceability throughout the software development lifecycle. The configuration items are generally documents in the early part of the development lifecycle, whereas the focus is on source code control management and software release management in the later parts of the development lifecycle.

The company standards need to be adhered to, and the correct version of a work product should be known at all time. There is a need for a document and source code repository, which has access controls, checking in and checking out procedures; and labelling of releases.

A project will have a configuration management plan, and the configuration manager role is responsible for ensuring that the configuration management activities are carried out correctly.

Configuration management ensures that the impacts of proposed changes are considered prior to authorization. It ensures that releases are planned, and that only authorized changes to the software are made. The integrity of the system is maintained, and the constituents of the software system and their version numbers are known at all times. Configuration audits will be conducted to verify that the CM activities have been carried out correctly.

# Epilogue

# 16

We embarked on a long journey in this book and set ourselves the objective of providing a concise introduction to the software testing field for students and practitioners. The book was based on the author's industrial experience in the software quality and process improvement fields, and it covered both theory and practice. The objective was to give the reader a grasp of the fundamentals of the software testing field, as well as guidance on how to apply the theory in an industrial environment.

Customers today have very high expectations on quality, and expect high-quality software to be consistently delivered on time and on budget. This requires sound software engineering practices be employed to enable quality software to be consistently produced. Further, the quality of the delivered software is closely related to the quality of the underlying processes used to build the software, and on adherence to them.

Many processes are employed in the development and testing of software, and companies need to determine the extent to which the processes are fit for purpose. The process will need to be continuously improved and often model-based improvement frameworks are employed. Piloting or technology transfer of innovative technology is an important part of continuous improvement. Companies need to ensure that the desired quality is built into the software product.

Chapter 1 discussed the fundamentals of software quality field including a history of quality and the pioneers that have influenced the field. We discussed the work of key figures such as Deming, Juran, and Crosby, as well as the work of Watts Humphrey who is considered the father of software quality.

Chapter 2 presented a broad overview of the fundamentals of software engineering, and we discussed various software lifecycles models. We discussed the various activities in the waterfall model, including requirements gathering and specification, software design, implementation, testing, and maintenance. The lightweight Agile methodology was discussed, and it has become very popular in the software engineering field.



G. O'Regan, *Concise Guide to Software Testing*,
Undergraduate Topics in Computer Science,
https://doi.org/10.1007/978-3-030-28494-7_16

Chapter 3 provided an introduction to the fundamentals of software testing in traditional software engineering, and we discussed the various types of testing that may be carried out during the project. We discussed test planning, test case definition, test environment set-up, test execution, test tracking, test metrics, test reporting, and testing in an e-commerce environment.

Chapter 4 discussed static testing, which plays an important role in building quality into a product. We discussed the well-known Fagan inspection process that was developed at IBM in the 1970s, as well as lighter review and walkthrough methodologies. We discussed the static code analysis of software code without executing the code, which is usually performed with automated tools.

Chapter 5 discussed software test planning, where testing is a subproject of a project and needs to be managed as such. Test planning involves defining the scope of the testing to be performed; defining the test environment; estimating the effort required to define the test cases and to perform the testing; identifying the resources needed (including people, hardware, software, and tools); assigning the resources to the tasks; defining the schedule; and identifying any risks to the testing and managing them.

Chapter 6 discussed test analysis and design, which is concerned with determining the test conditions, and designing the test cases (using various techniques) for the testing. The requirements and test conditions are used to specify the test cases, where each test case includes input, the test procedure for carrying out the test, and the expected results.

Chapter 7 discussed test management, which is concerned with the activities involved in managing the software testing process. Chapter 8 is concerned with test outsourcing and is concerned with the selection and management of a testing supplier. We discussed how candidate suppliers are identified, formally evaluated against defined selection criteria, and how the appropriate supplier is selected. We discussed how the selected supplier is managed during the testing.

Chapter 9 is concerned with test metrics and problem-solving, and includes a discussion of the Goal, Question, Metrics (GQM) approach. GQM allows appropriate metrics related to the organization goals to be defined. A selection of sample metrics is presented, and problem-solving tools such as fishbone diagrams, Pareto charts, and trend charts were discussed.

Chapter 10 discussed various tools to support the various activities in software testing. The focus is first to define the process, and then to find tools to support the process. Tools to support test management were discussed, as well as tools to support test design and execution, static testing, performance and monitoring tools, and defect tracking tools.

Chapter 11 discussed test process improvement. It began with a discussion of software processes, and discussed software process improvement initiatives. We discussed dedicated test process improvement models such as the TMM, TPI, TMap, STEP, and CTP.

Chapter 12 discussed the Agile methodology and software testing. Agile is a popular lightweight approach to software development, and it provides opportunities to assess the direction of a project throughout the development lifecycle.

Ongoing changes to requirements are considered normal in the Agile world, and Agile has a strong collaborative style of working.

Chapter 13 discussed the verification of safety-critical systems, where such systems often need an extra level of assurance in their correctness. Formal methods consist of a set of mathematical techniques that support the development and verification of safety-critical systems. They may be employed to provide a rigorous proof that the implemented program satisfies its specification, and they have been mainly applied to the safety-critical field.

Chapter 14 discussed legal and ethical aspects of software testing, and we discussed professional responsibility and the ethics of the professional tester. We discussed legal aspects of computing including intellectual property law such as patents, copyright, trademarks, and trade secrets. We discussed bespoke software development and test outsourcing and the licensing of software. We discussed professional negligence in the development and testing of software and computer crime, and legal aspects of failure including lawsuits and the law of tort.

Chapter 15 discussed software configuration management including the fundamental concept of a baseline. Configuration management is concerned with identifying those deliverables that must be subject to change control and controlling changes to them. This chapter is the concluding chapter in which we summarized the journey that we have travelled in this book.

## 16.1 The Future of Software Testing

Software testing has come a long way since the 1950s and 1960s, when it was accepted that the completed software would always contain lots of defects, and that the coding should be done as quickly as possible, to enable these defects to be quickly identified and corrected. The software crisis in the late 1960s highlighted problems with the quality and reliability of the delivered software and led to the birth of software engineering as a discipline in its own right.

The software engineering field is highly innovative and many new technologies and systems have been developed over the decades. These include object-oriented design and development; formal methods and UML; the waterfall and spiral models; software inspections; software testing; software process improvement; the CMMI; and the Agile methodology.

Software testing will continue to be fundamental to the success of projects. There is not a one size that fits all: some companies (e.g. in the safety-critical or security-critical fields) are likely to focus on more rigorous techniques such as formal methods and software process maturity models such as the CMMI. For other domains, the lightweight Agile methodology with its test-driven development may be the appropriate approach.

Companies are likely to measure the cost of poor quality in the future, as driving down the cost of poor quality will become more important. Software components and the verification of software components is likely to become important, in order to speed up software development and to shorten time to market. Software reuse and open source software development is likely to grow in popularity, and continuous innovation will continue in the software engineering and testing fields.

# Glossary

**ACM** Association for Computing Machinery

**AQL** Acceptable Quality Level

**ATM** Automated Teller Machine

**BCS** British Computer Society

**CBA IPI** CMM Based Assessment Internal Process Improvement

**CBA SCE** CMM Based Assessment Software Capability Evaluation

**CCB** Change Control Board

**CEI** Computer Ethics Institute

**CM** Configuration Management

**CMM®** Capability Maturity Model

**CMMI®** Capability Maturity Model Integration

**COCOMO** Constructive Cost Model

**COPQ** Cost of Poor Quality

**COTS** Customized Off the Shelf

**CR** Change Request

**CTP** Critical Test Process

**DOD** Department of Defence

**DOORS** Dynamic Object-Oriented Requirements System

**DSDM** Dynamic Systems Development Method

**ESA** European Space Agency

**EULA** End User License Agreement

**FMEA** Failure Mode Effects Analysis



G. O'Regan, *Concise Guide to Software Testing*,
Undergraduate Topics in Computer Science,
https://doi.org/10.1007/978-3-030-28494-7

**FSF** Free Software Foundation

**GNU** GNU's Not Unix!

**GPL** General Public License

**GQM** Goal, Question, Metric

**GUI** Graphical User Interface

**HP** Hewlett Packard

**HR** Human Resources

**HTML** Hyper Text Mark-up Language

**IBM** International Business Machines

**IDE** Integrated Development Environment

**IEC** International Electro technical Commission

**IEEE** Institute of Electrical and Electronic Engineers

**ISEB** Information System Examination Board

**ISO** International Standards Organization

**ISTQB** International Software Testing Qualification Board

**JAD** Joint Application for Development

**KLOC** Thousand Lines of Code

**KPI** Key Performance Indicator

**LCL** Lower Control Limit

**LDRA** Liverpool Data Research Associates

**LOC** Lines of Code

**MTBF** Mean Time Between Failure

**MTP** Master Test Plan

**MTTF** Mean Time to Failure

**MTTR** Mean Time to Repair

**NATO** North Atlantic Treaty Organization

**ODC** Orthogonal Defect Classification

**OJEU** Official Journal of European Union

**OOD** Object-Oriented Design

**PBI** Product Backlog Item

**PCE** Phase Containment Effectiveness

**PDCA** Plan, Do, Check, Act

**PMBOK** Project Management Book of Knowledge

**PMI** Project Management Institute

**PMP** Project Management Professional

**PRINCE** Projects In a Controlled Environment

**PSP** Personal Software Process

**PVCS** Polytron Version Control System

**QCC** Quality Control Circle

**QTP** Quick Test Professional

**RAD** Rapid Application Development

**RAG** Red, Amber, Green

**RFP** Request for Proposal

**ROI** Return on Investment

**RUP** Rational Unified Process

**SCAMPI** Standard CMMI Appraisal Method for Process Improvement

**SCM** Software Configuration Management

**SEI** Software Engineering Institute

**SG** Specific Goal

**SLA** Service Level Agreement

**SLOC** Source lines of code

**SOW** Statement of Work

**SP** Specific Practice

**SPC** Statistical Process Control

**SPI** Software Process Improvement

**SPICE** Software Process Improvement Capability dEtermination

**STEP** Systematic test and evaluation process

**TDD** Test Driven Development

**TMap** Test Management Approach

**TMM** Test Maturity Model

**TPI** Test Process Improvement

**TQM** Total Quality Management

**TSP** Team Software Process

**UAT** User Acceptance Testing

**UCL** Upper Control Limit

**UFT** Unified Functional Testing

**UK** United Kingdom

**UML** Unified Modelling Language

**URS** User Requirements Specification

**VDM** Vienna Development Method

**VOB** Version Object Base

**VSS** Visual Source Safe

**XP** Extreme Programming

**Y2K** Year 2000

**ZD** Zero Defects

# Index

**A**
Agile development, 44
Agile test principles, 231
Analogy method, 104
Appraisal, 26, 208
Ariane 5, 4
Ariane 5 disaster, 39
Automated software inspections, 94

**B**
Barriers to success, 207
Baseline, 273
Bespoke software, 261
Black box testing, 120
Boundary value analysis, 122
Breakthrough and Control, 10
BugDigger, 196
Bugzilla, 194
Business ethics, 252

**C**
Capability Maturity Model Integration (CMMI), 204
Capture/playback tools, 193
Change control, 278
Change control board, 111
Change request, 111
Cleanroom, 238
Cleanroom methodology, 244
Clearcase, 273
ClearQuest, 273
CMMI maturity model, 53
CMMI model, 217
Computer ethics, 254
Configuration control, 274
Configuration identification, 274
Configuration management, 115, 271
Configuration management audits, 279
Configuration management plan, 277
Constructive Cost Model (COCOMO), 186, 198
Corporate social responsibility, 253
Cost of poor quality, 22, 82, 166
Cost predictor models, 104
CTP model, 216
Customer care metrics, 164
Customer satisfaction, 25
Customer satisfaction metrics, 157

**D**
Data gathering for metrics, 168
Debugging tools, 192
Decision table, 122
Decision testing, 126
Defect tracking tools, 194, 196–198, 284
Defect-type, 92
Delphi method, 104
Document control management, 276
Dynamic Object-Oriented Requirements System (DOORS), 189, 190

**E**
E-Commerce testing, 72
Edwards Deming, W., 8
Equivalence partitioning, 121
Error guessing, 127
Escrow agreement, 150, 262
Estimation, 102
Estimation in agile, 227
Ethical software tester, 255
Ethics, 251
European Space Agency (ESA), 39
Experienced based testing, 127
Expert judgment, 104
Exploratory testing, 127

**F**
Fagan inspection guidelines, 88


G. O'Regan, *Concise Guide to Software Testing*,
Undergraduate Topics in Computer Science,
https://doi.org/10.1007/978-3-030-28494-7

Fagan inspections, 37, 86
Fishbone diagram, 21, 171
Formal methods, 54, 242
Formal methods and testing, 245
Formal specification, 242
Fred Brooks, 5
Functional requirement, 119
Function points, 104

G
Goal Question Metric (GQM), 24, 154

H
Hacker, 265
Histogram, 21, 172
HP quality center, 185
HP Unified functional testing, 193

I
IBM Rational ClearQuest, 195
Identifying suppliers, 147
IEEE standards, 41
Inspection meeting, 90
Integrity tool, 191
ISO 9001, 204
ISO 9126, 5

J
Jira, 194
Joseph Juran, 10
Jtest, 188

L
Law of Tort, 260
Lawsuits and professional negligence, 259
LDRA Tool, 188
Legal aspects of testing, 257
Legal aspects of test outsourcing, 261
Legal impacts of failure, 259
LoadRunner, 195

M
Maintenance, 51
Managing change requests, 139
Managing defects, 137
Measurement, 153
Michael Fagan, 19
Microsoft project, 187
Model, 41
Model checking, 247
Model checking and testing, 247
Mongolian Hordes Approach, 33

N
Non-functional requirement, 119

P
Pair programming, 228
Pareto chart, 22, 173
Parnas, 37, 48, 256
PDCA model, 217
Performance testing, 50
Performance test tools, 195
Personal Software Process (PSP), 17, 24, 204, 205
Phase containment effectiveness, 94
Philip Crosby, 5, 12
Project Management Book of Knowledge (PMBOK), 101
Polytron Version Control System (PVCS), 197, 273
Prince 2, 37, 101
Problem-solving techniques, 169
Process mapping, 205
Process model, 204
Professional Engineering Association, 35
Professional engineers, 38, 257
Professional ethics, 252
Professional responsibility, 257
Project board, 101, 112
Project closure, 114
Project management, 52
Project management metrics, 158
Project manager, 101
Project monitoring and control, 140
Prototyping, 47, 119
Psychology of software tester, 70
PV Tracker, 194

Q
QA Complete, 186
QTest, 186

R
Rational unified process, 41, 43
Refinement, 243
Request for Proposal (RFP), 148
Requirements engineering, 119
Requirements validation, 118, 243
Requirements verification, 118
Requirement traceability, 74, 130, 132
Risk management, 107, 141

S
Safety critical system, 236
Scatter graphs, 176

Scrum methodology, 225
Security, 267
Selenium, 193
Shewhart model, 7
Silk performer, 195
Silk Test, 194
Six sigma, 52
Software crisis, 34, 55
Software defect, 68
Software dependability, 240
Software design, 118
Software engineering, 34, 36, 285
Software failures, 39
Software inspections, 79
Software licensing, 263
Software process, 202
Software process improvement, 23
Software reliability, 237, 239
Software reuse, 49
Software testing, 49, 59, 60
Software testing in agile, 229
Software testing tools, 181
Source code control management, 277
Spiral model, 41
Sprint planning, 45, 223
Standish group, 2, 35
Statement coverage, 126
Statement of work, 150
State transition testing, 123
Static testing, 79
Statistical Process Control (SPC), 7, 176
Statistical quality control, 7
Statistical usage testing, 245
STEP model, 215
Story, 45, 223
Structured walkthrough, 83
System testing, 50

**T**
Tacoma narrows bridge, 3
Taurus project, 2
Team Software Process (TSP), 205
Test case design, 66
Test cases, 63
Test case specification, 128
Test completion, 143
Test conditions, 120
Test coverage tools, 192
Test design, 120
Test Driven Development (TDD), 50, 72, 230, 231
Test environment, 63
Test execution, 67, 136
Test execution metrics, 159
Test execution tools, 191
Test harness tools, 192
Testing and computer crime, 264
Testing and hacking, 265
Test level, 63
Test management tools, 184
Test monitoring and control, 110, 140
Test outsourcing, 145
Test planning, 62, 65
Test process, 61
Test process improvement, 199
TestRail, 186
Test reporting, 67, 112, 141
Test status, 65
Test team, 135
Test tools, 65, 75
TMap next model, 212
TMM*i* model, 211
Tom Gilb, 24
Total quality management, 6
TPI Next model, 214
Traceability, 48
Trend graph, 22, 175

**U**
UML and testing, 246
Unit testing, 49
Use-case diagram, 125
Use case testing, 124
User requirements, 118, 131
User stories, 226

**V**
Victor Basili, 24, 155
Vienna Development Method (VDM, 244
Visual Source Safe (VSS), 273

**W**
Walter Shewhart, 7, 200
Waterfall model, 41
Watt Humphrey, 16
White box testing, 125
Work breakdown structure, 104

**Y**
Y2K, 35
Y2K bug, 3, 39

**Z**
Z, 244

CPSIA information can be obtained
at www.ICGtesting.com
Printed in the USA
LVHW050103070220
646091LV00002B/18